Advances in Experimental Medicine and Biology

Volume 1146

AF739049

More information about this series at http://www.springer.com/series/5584

Caterina A. M. La Porta • Stefano Zapperi
Editors

Cell Migrations: Causes and Functions

Springer

Editors
Caterina A. M. La Porta
Department of Environmental Science and Policy,
Center for Complexity and Biosystems
University of Milan
Milan, Italy

Stefano Zapperi
Department of Physics,
Center for Complexity and Biosystems
University of Milan
Milan, Italy

ISSN 0065-2598 ISSN 2214-8019 (electronic)
Advances in Experimental Medicine and Biology
ISBN 978-3-030-17595-5 ISBN 978-3-030-17593-1 (eBook)
https://doi.org/10.1007/978-3-030-17593-1

This Springer imprint is published by the registered company Springer Nature Switzerland AG.
The registered company address is: Gewerbestrasse 11, 6330 Cham, Switzerland

Preface

Cell migration is a key ingredient of organized life forms. It plays a key role both during embryo development and during adult life where is important in many physiological processes, such as the immune response or wound healing, as well as in pathological conditions like cancer. Improvements in microscopy techniques together with the possibility to carry out quantitative analysis by tracking cells in vitro or in vivo using animal models have opened a new perspective on this delicate and fundamental process.

The present book reviews the most recent and innovative approaches on cell migration with the aim to obtain a complete picture of this complex process. In our opinion, the most interesting recent approaches on cell migration make use of an interdisciplinary perspective, combining biology with biophysics and bioengineering which provide new experimental techniques and devices, as well as quantitative image analysis tools. Furthermore, our understanding of cell migration relies more and more on the definition of theoretical concepts and numerical simulations, all of which are reviewed in depth in this book.

The intended readership for this book is provided by graduate students in biology and biophysics who are interested to reach a deeper understanding of the new aspects of cell migration and more experienced researchers who are new in this field. The book can also be read by selecting individual chapters that are always self-contained and up to date.

The book starts with the chapter of Yang et al. (Chap. 1) discussing the most important computational models of collective cell migration. The authors consider mechanical models where cells and extracellular matrix interact through physical forces. At the same time, the authors discuss the underlying biochemical models where rate equations describe molecular interactions at many regulatory levels. The general goal is to combine physical and biochemical aspects into a unique framework. The authors also discuss an important issue playing a role in physiological and pathological processes: the transition from mesenchymal individual migration to collective motion and its relation to the traditional epithelial-mesenchymal transition.

Chapter 2, by Luca Giomi, presents a detailed account of contour models describing the complex interaction between cells and the extracellular matrix that is essential for cell migration. This area of activity has received a wide attention in recent years, thanks to the development of traction force microscopy. This technique allows to measure directly mechanical properties of the living cells as they adhere or crawl on micro-patterned substrates.

Ferrari and Giampietro (Chap. 3) discuss the most promising experimental models and techniques to study epithelial cells and focus also on a related cellular compartment: the endothelium. These cells are involved in many physiological processes such as wound healing, but changes in their arrangements could help in pathological process such as tumor spreading. The authors discuss the forces generated by these cells and the most advanced experimental techniques to measure them. Banerjee and Marchetti's chapter (Chap. 4) focuses instead on theoretical models of cell aggregates seen as continuous active material. This analogy provides powerful mathematical tools to describe the physiological dynamics that underlie developmental process where the cells move into a viscous fluid and stiffen to support mechanical stresses and maintain cohesion.

Statistical analysis helps highlight collective properties of cell migration, properties that turn out to be universal and independent on cell types or the form of the extracellular matrix. These aspects are discussed in the chapter by La Porta and Zapperi (Chap. 5) who also highlight similarities and differences between cell assemblies and soft matter systems.

In Chap. 6, the authors discuss how the development of new technologies such as microfluidic devices plays an important role in obtaining single-cell and single-molecule information. This approach provides an important tool to study confinement-induced cell migration and investigate how the environment can help or hinder cell migration. In particular, the capability of the cell to organize invadosomes under confinement contributing actively to its migration and invasion is a topical aspect of this field. It is important to always bear in mind the dual characteristics of cells: the collective migration and the capability of single cells to move in response to an external stimulus that can be chemical or physical, due to the contact between cells. Microfluidic technologies are bringing this research area to a higher level of complexity and can really boost our understanding of the interaction of cells and the microenvironment.

Chapter 7 by Linus Schumacher discusses collective cell migration in development and in particular in *Drosophila*. The main features of this process include the heterogeneity of cell states, substrate-free migration, contact-inhibition of locomotion, and reprogramming. The possibility to disentangle the main principles that underly cellular migration during development is important to understand pathologies occurring during development but also cancer. In this connection, Chap. 8 considers nuclear mechanics in cancer migration in relation to genome integrity and potentially cancerous mutations.

Milan, Italy Caterina A. M. La Porta
Milan, Italy Stefano Zapperi

Contents

Computational Modeling of Collective Cell Migration: Mechanical and Biochemical Aspects

1

Yanjun Yang, Mohit Kumar Jolly, and Herbert Levine

Abstract

Collective cell migration plays key roles in various physiological and pathological processes in multicellular organisms, including embryonic development, wound healing, and formation of cancer metastases. Such collective migration involves complex crosstalk among cells and their environment at both biochemical and mechanical levels. Here, we review various computational modeling strategies that have been helpful in decoding the dynamics of collective cell migration. Most of such attempts have focused either aspect – mechanical or biochemical regulation of collective cell migration, and have yielded complementary insights. Finally, we suggest some possible ways to integrate these models to gain a more comprehensive understanding of collective cell migration.

Keywords

Collective cell migration · Epithelial mesenchymal plasticity · Subcelluar element model · Phase field model · Wound healing

Y. Yang · H. Levine (✉)
Center for Theoretical Biological Physics, Rice University, Houston, TX, USA
e-mail: herbert.Levine@rice.edu

M. K. Jolly
Centre for BioSystems Science and Engineering, Indian Institute of Science, Bangalore, India

1.1 Introduction

Cells are basic building blocks for biological organisms and their migration plays an essential role in these systems. In many scenarios, cells move collectively instead of moving as single entities. Collective cellular migration is central to many physiological process in multicellular organisms, including embryonic development, wound healing, and metastasis formation. The mechanisms for this collective cellular migration are complex, involving mechanical and biochemical interactions between cells and their environment. Depending on the biochemical and mechanical environment as well as specific biological context, collective migration may involve different mechanisms (examples shown in Fig. 1.1). In this chapter, we introduce basic mechanical and biochemical models for collective cell migration.

Single-cell movement is the simplest form of cell migration. *In vivo*, two typical types of single-cell movements discussed are amoeboid movement and mesenchymal movement. Such single-cell movements result from weak or unstable intercellular junctions and occur under certain extracellular tissue conditions. On the other hand, in collective cellular migration, stronger intercellular interactions exist at mechanical and/or biochemical levels (te Boekhorst et al. 2016). What migration mode a cell adopts is controlled

C. A. M. La Porta, S. Zapperi (eds.), *Cell Migrations: Causes and Functions*, Advances in Experimental Medicine and Biology 1146,
https://doi.org/10.1007/978-3-030-17593-1_1

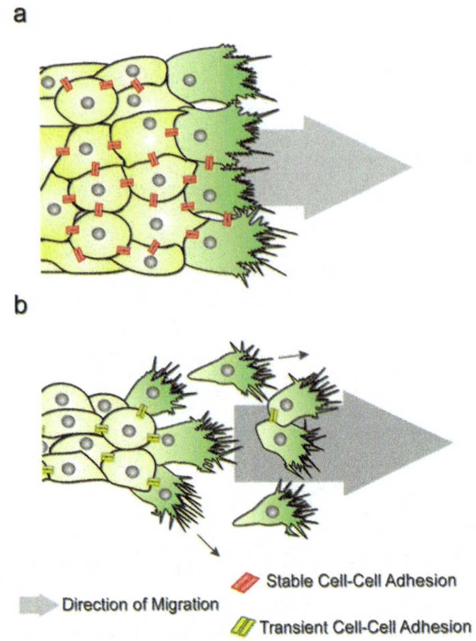

Fig. 1.1 Examples for collective cell migration. (Scarpa and Mayor 2016). (**a**) Epithelial Collective migration (**b**) Mesenchymal Collective Migration. (https://www.ncbi.nlm.nih.gov/pmc/articles/PMC4738384/figure/fig1/)

by both local and global mechanical and biochemical signals and their complex crosstalk. So far, the mechanical and biochemical aspects of cell migration have mostly been studied independently, both in experiments and computational modeling. However, to accurately decipher the mechanism for collective cell migration, we need to consider them in an integrated fashion. For example, in collective to individual transitions, biochemical signals change the mechanical properties of the cells, while, on the other hand, the change in mechanical properties feeds back to the biochemical signaling. In these transitions, mechanical properties change, such as a down-regulation of adherens junctions formation (Haeger et al. 2015). These changes are often related to biochemical signals such as Rac, Rho, Integrin etc., and they, in turn, often induce changes in mechanical properties.

1.2 Mechanical Models

Modeling both mechanical and biochemical aspects coupled together is quite complex and there are only a few examples of this type of complete treatment. Therefore, we start with several purely mechanical models. These models simulate the migration modes and force patterns based on the mechanical properties including forces within and among cells, interactions between cells and extracellular matrix. It is worth remembering that a good model should only contain necessary factors or parameters; it should not be too complex trying to simulate everything, thus leading to overfitting and reducing the confidence in the ability of the model to predict novel behaviors. In other words, we should create a model capturing the key factors for the phenomena that we are interested in and consider their relationships. In a model for collective cell migration, we need to decide how and to what degree to specify the following key elements: (1) the motility of single cells within the collective, (2) the cell shape and intercellular interactions, (3) the mechanism for a cell to choose its moving direction, (4) potential biochemical signaling (Camley and Rappel 2017). Here, we discuss various frameworks that have been adopted to model collective cell behavior.

1.2.1 Agent-Based Models

An agent-based model is the simplest scheme for representing collective cell migration, where every cell is treated as a single particle (Fig. 1.2a). In this type of model, intercellular interactions can be incorporated in a simple manner, while still being able to capture some macroscopic features. Usually, we can treat the cell-cell adhesion and cell-cell volume exclusion as an attraction or repulsion force between agents. For cell-substrate interaction and other cell-cell interactions, we can treat them as friction that a cell faces during migration. To obtain collective movement from this approach, usually a Vicsek-like model is adopted (Vicsek et al. 1995). That is, it is assumed that collective motion is due to alignment of the self-propulsion among the agents in this model, taking place via a variety of mechanisms such as cell-cell alignment, velocity alignment etc. In the original Vicsek formulation, every agent has a self-propulsion with a constant speed v and a changing direction of the velocity

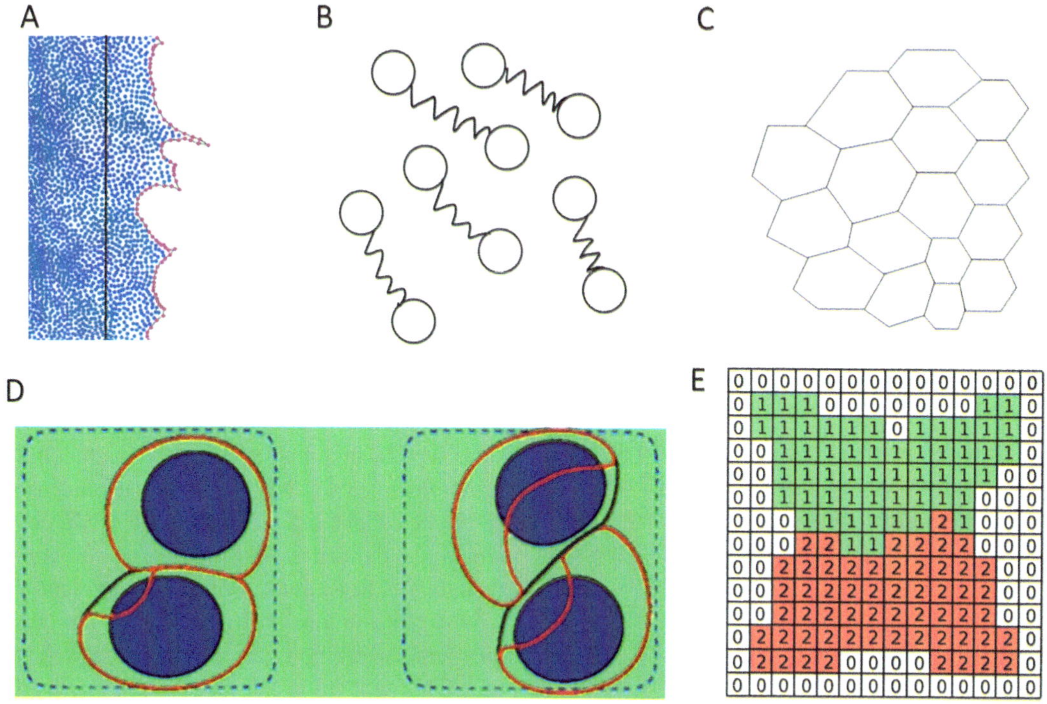

Fig. 1.2 Example figures for different mechanical models. (**a**) An agent-based model simulating an expanding cell layer (Tarle et al. 2017) (http://iopscience.iop.org/1478-3975/14/3/035001/downloadHRFigure/figure/pbaa6591f01) (**b**) A cartoon for a two-subcellular-element model. (**c**) A cartoon for a Vertex/Voronoi model (**d**) A phase field model simulating a pair of rotation cells (Camley et al. 2014) (http://www.pnas.org/content/pnas/111/41/14770/F1.large.jpg). (**e**) A cellular Potts model (Albert and Schwarz 2016) (https://journals.plos.org/ploscompbiol/article/figure/image?size=original&download=&id=10.1371/journal.pcbi.1004863.g002)

θ. At each simulation time step, every agent i first aligns its θ with that of its neighbors, by taking the average θ of its neighbors plus an uncertainty; then we update the position of agent i using the new velocity. Specifically, we update the θ in Vicsek model as described in Eq. (1.1),

$$\Theta_i\left(t+\Delta t\right)=\left\langle\Theta_j\right\rangle_{nearest\ neighbors\ of\ i}+\eta_i(t). \tag{1.1}$$

This kind of agent-based model is sufficient to capture some key features in collective cellular motion. For example, Camley et al. studied emergent collective chemotaxis in the absence of single-cell gradient sensing (Camley et al. 2016). To accomplish this, they expanded the agent-based model by introducing a chemical gradient for the cell cluster, a polarity for each cell, and contact inhibition of locomotion between cells. They provided a quantitative minimal model for collective chemotaxis without the need for incorporating single-cell chemotaxis. Another example of agent-based model in the context of collective migration aimed to explain the emergence of finger-like protrusion at wound frontiers (Sepúlveda et al. 2013). Here, an Ornstein-Uhlenbeck process was to drive the cell, a linear damping term to simulate the adhesive contacts or friction with the substrate, cell-cell interactions and a velocity alignment with each cell's nearest neighbors. Aside from these basic settings, by assigning leader cells which moves faster at the front edge of the wound, they successfully reproduced the finger-like protrusions observed experimentally (Poujade et al. 2007; Petitjean et al. 2010; Reffay et al. 2011, 2014). In particular, the velocity alignment gives rise to a velocity gradient from the leading front to the back end of the cellular sheet. This velocity gradient induces the instability at the wound front and forms the "fingers". This model is highly adaptable in which we can easily include additional intercellular features, such as contact inhibition of locomotion, different kind of intercellular forces based on cell type or position etc. For example, in this model, a contour force at the wound edge can be added (Tarle et al. 2015), that can empirically simulate long distance intercellular tensions on the wound rim.

1.2.2 Sub-Cellular-Element Model

With only one point-particle to represent a cell, the agent-based approach can sometimes be too simple. The most significant drawback of this model is that it cannot take into account detailed intracellular features including actin-myosin contractions, cell morphology, cell division etc. To resolve this issue, one can turn to a subcellular-element model, where each cell is represented by several subcellular-elements (Sandersius and Newman 2008; Sandersius et al. 2011). In the subcellular-element model, in addition to everything included in the agent-based model, other factors such as intracellular contractility, cell polarization, cell size, and cell division can be included as well. The simplest subcellular-element model is the two-subcellular-element model, where two subcellular-elements are used to represent a cell, one for the front end and another one for the back end (Zimmermann et al. 2016) (Fig. 1.2b).

This two-particle-based model can be validated by experiments in several ways. For example, we can experimentally measure the morphology of the collective cell culture, the position and velocity of the cells, and the traction force between cells and the substrate. These mechanical properties can be compared with the force patterns and velocity fields generated from the simulations of the mathematical model. Using a two-subcellular-element model, our group successfully deciphered the role of the supracellular actomyosin cable around the wound (Yang and Levine 2018), in the context of wound healing. To do this, we considered a self-propelled force m which is subject to contact inhibition of locomotion (Zimmermann et al. 2016) for each subcellular-element, an intracellular contractive force between two subcellular-element f_{contr} and an intercellular interaction between elements from different cells $f_{rep/adh}$ as shown in Eq. (1.2),

$$v = \frac{1}{\xi}\left(m + f_{contr} + f_{rep/adh}\right). \qquad (1.2)$$

We updated the position for each subcellular element by a simple Euler scheme $x = v\ dt$. The traction force can be calculated by $m - \xi v$, where ξ is the friction coefficient between the cell and substrate. To take into account the supracellular actomyosin cable, we add a set of mechanical links connecting elements from cable cells, which are defined as those cells on the wound boundary. These links form a ring around the wound and the tension from the ring simulates the purse-string contraction of the actomyosin ring. This simulation can successfully predict the traction force patterns during the wound healing process under different conditions. By comparing these force patterns with both experiments and their corresponding simulations (Brugués et al. 2014; Vedula et al. 2015; Martin et al. 2015), we successfully determined how the actomyosin ring around the wound can contribute to the driving of this collective movement. More importantly, we validated this strategy for modeling a supracellular actomyosin cable. This framework can be applied in many other scenarios, including the intercellular cable on both sides of fingering-like protrusions. Thus, an overall advantage of the particle-based formulation is that it is powerful both conceptually and computationally but still simple enough to enable us to simulate a large number of cells while being able to reasonably mimic both intracellular and intercellular properties for each cell.

1.2.3 Vertex/Voronoi Model

The above discussed models do not specifically consider cell morphology. To better take into consideration the cell morphology, especially for the case of epithelial cell layers, vertex and voronoi models assume that cells can be approximated with polygonal shapes (Fig. 1.2c). In the vertex model, a cell is parameterized by a set of vertices that mark the common point of three or more neighboring cells. On the other hand, in the Voronoi model, the common borders of neighboring cells are determined by the Voronoi construction (Honda 1978), i.e. a cell is defined by its center and any point within the region of this cell is closer to this cell's center than any other cell's center. Thus, a Voronoi diagram is similar to the Wigner-Seitz-cell in solid state physics (Wigner and Seitz 1933). In these models, the cell shape itself, the adhesion between cells and many other aspects related to the cell shape and boundary can be treated in a more accurate way. To investigate the collective behavior, a term for mechanical energy for each cell is included. This term is related to a cell's area and perimeter, and further information regarding cell-cell adhesion can be obtained from this energy. In addition, it is possible to include self-propulsion and even traction force in this approach. Bi et al. applied a self-propelled Voronoi model to demonstrate a jamming transition from a solid-like state to a fluid-like state in a confluent tissue (Bi et al. 2015, 2016). They write down the total energy of the system based on the area and perimeter of each cell, which are completely determined by the positions of the center of each Voronoi cell r_i. As shown in Eq. (1.3),

$$E = \sum_{i=1}^{N} \left[K_A \left(A\left(r_i\right) - A_0 \right)^2 + K_P \left(P\left(r_i\right) - P_0 \right)^2 \right], \tag{1.3}$$

A_{r_i}, P_{r_i} are the area and perimeter for cell at r_i, A_0, P_0 are the preferred values, and K_A, K_P are the area and perimeter moduli (Bi et al. 2016). This energy corresponds to cell volume incompressibility, contractility of the actomyosin cortex, and cell membrane tension and cell-cell adhesion. They show that the solid to fluid transition is only related to the single-cell moving speed, the persistence time of a cell track, and a shape index related to the area and perimeter of the cell. Their results provide a framework to understand this collective solid-to-liquid transition and a way to simulate cells with a defined shape. More recently, Koride et al. combined the Vertex model with biochemical regulation of contractility to model collective cellular migration in confluent epithelia (Koride et al. 2018). They incorporated cell-substrate friction, cell-cell friction, passive force and active force including contractility in the vertex model. They also included a Rho-ROCK-myosin signaling pathway to regulate the contractility, providing an example which combines mechanical and biochemical traits.

1.2.4 Phase Field Model

A phase-field model is a more elaborate construction which defines the cell boundary based on a set of partial differential equations for the phase field ϕ (Shao et al. 2010, 2012; Ziebert et al. 2012) (Fig. 1.2d). For a single cell, we often start with energy, including surface energy which is proportional to the cell's perimeter, bending energy etc. We can have the Hamiltonian for a single cell in a form similar to Eq. (1.4):

$$\mathrm{H} = \gamma \int d^2 r \left[\frac{\epsilon}{2} \left| \nabla \varnothing \right|^2 + \frac{G(\phi)}{\epsilon} \right] + \kappa \int d^2 r \, \frac{1}{2\epsilon} \left[\epsilon \, \nabla^2 \phi - \frac{G'(\phi)}{\epsilon} \right]^2, \tag{1.4}$$

where the first term is related to the cell perimeter and second term is the energy for the curvature integrated over membrane. Such models typically apply a constraint that the area of the cell is conserved during migration and deformation, but the perimeter of the cell can vary. Based on this information, we can then write down expressions for the forces in the system and establish the equation for the evolution of the phase field. In one popular version, the forces are generated by the cytoskeleton acting as an active gel (Prost et al. 2015) with separate terms corresponding to polymerization and myosin-based contraction. Within this model, we can simulate many details in a cell such as actin dynamics, membrane bending, protrusion, focal adhesion etc.

For multiple cells, each cell is represented by a phase field ϕ^i. In this context, we need to also consider intercellular interactions which can be given by an additional term in the Hamiltonian. Using these ideas, Camley et al. studied the rotation of a pair of mammalian cells (Camley et al. 2014) that can be experimentally observed (Segerer et al. 2015). They investigated the effect of various cell polarity mechanisms on the rotation motion, including contact inhibition of locomotion, alignment of position or velocity with neighboring cells. They show that the velocity alignment promotes the persistent rotational motion robustly. For simulation of more cells, Löber et al. provide an alternative phase-field model can simulate hundreds of cells (Löber et al. 2015).

1.2.5 Cellular Potts Model

Another methodology, called the cellular Potts model, borrows the idea of spin from physics. This model represents each cell as the set of points on a lattice all having the same spin, and so these N cells give rise to a N-state model (Fig. 1.2e). We evolve the system by flipping the spin value at a random site on the lattice with a randomly chosen probability, and decide whether to accept the flipping based on the Hamiltonian, which often needs to be minimized. The Hamiltonian often contains adhesion energies, cell volume exclusion etc. It is determined by the configuration of the lattice. It can be written in forms of the spin values as shown in Eq. (1.5),

$$H = \sum_{i,j\ neighbors} J_{ij}\left(1 - \delta_{\sigma(i),\sigma(j)}\right) + \sum_i \lambda\left(A_i - A_{i,0}\right)^2. \tag{1.5}$$

According to this equation, we can flip the spin on any site and hence in principle cells can split apart. However, if the site is located in the middle of a cell, it the energy change will be prohibitively large and thus this flip will be rejected. We usually use a Metropolis algorithm to update the Hamiltonian. Using this model, we can simulate cell dynamics, including cell sorting, cell migration etc. (Graner and Glazier 1992; Szabó et al. 2010; Albert and Schwarz 2016). Using this Cellular Potts approach, Albert and Schwarz simulated collective behavior on adhesive micropatterns. By combining the Cellular Potts model with previous models developed for cell mechanics, cell division, cell migration, cell-cell adhesion, they generated a framework which can explain a large range of experiments (Albert and Schwarz 2016).

1.3 Biochemical Models

The abovementioned mechanical models can simulate collective cellular motions. They all have their own assumptions, characteristics, strengths and limitations; the choice of a model should be guided by the relevant question at hand, and typically depend on the tradeoff between what details should be included and how much resources (computation and manpower) are available. It is also possible for these models to be connected to the biochemical signaling aspect, but with the exception of work on agent-based systems, this is not often attempted. Here, we discuss on pure biochemical signaling models relevant for collective cell migration.

Biochemical models are typically formulated using ordinary or partial differential equations (ODEs, PDEs) to represent a temporal/spatiotemporal profiles of various signaling molecules at a cellular or tissue-level. These equations represent molecular interactions at many regulatory levels that have been experimentally reported, such as transcriptional or post-transcriptional regulation. However, detailed kinetic parameters for multiple interactions may not be readily available, especially as the network grows in size. Thus, an alternative approach typically adopts a parameter-free approach, only representing the interactions (inhibitory/excitatory) among a set of molecules – typically known as Boolean models.

In the context of collective cell migration, biochemical models have been constructed from the perspective of Epithelial-Mesenchymal Transition (EMT), a cell biological process that allows epithelial cells to weaken their cell-cell adhesion and lose their apico-basal polarity while gaining the mesenchymal traits of migration and invasion. Cells undergo EMT in multiple physiological and pathological contexts – gastrulation, branching morphogenesis and other aspects of embryonic development (type I EMT), wound healing and fibrosis (type II EMT), and cancer metastasis (type III EMT) (Kalluri and Weinberg 2009). EMT has been long considered to be a binary process; however, recent studies have demonstrated that cells may stably attain one or more hybrid epithelial/mesenchymal (E/M) phenotype(s) (Nieto et al. 2016; Jolly et al. 2015). These hybrid E/M phenotype(s) typically integrate adhesion (epithelial trait) and migration (mesenchymal trait), thus enabling collective cell migration, as observed during wound healing, branching morphogenesis, and more recently in cancer metastasis (Micalizzi et al. 2010; Cheung and Ewald 2016). Mouse models have shown that most metastases are formed by cells migrating collectively and eventually appearing as clusters of Circulating Tumor Cells (CTCs) in the blood system – typically 5–8 cells in size. Moreover, the presence of CTC clusters in patients may predict poor survival (Cheung and Ewald 2016).

Various biochemical signaling models have been proposed to gain insights how cells can attain a hybrid E/M phenotype. The first two models focus on interactions among two families of microRNAs miR-34 and miR-200, and two families of transcription factors ZEB and SNAIL. These models capture the kinetics of individual biochemical reactions using estimated parameters from the relevant experimental literature. Both the models predict that this network can enable the existence of a hybrid E/M or partial EMT phenotype (Fig. 1.3a), although their predicted expression signature for a hybrid E/M signature is somewhat different due to variations in assumptions (Bocci et al. 2018) – for instance, Lu et al. developed a specific microRNA-based framework to capture the mRNA-microRNA interactions (Lu et al. 2013). Experimental support for both these models has been presented, suggesting that different cell lines may exhibit different hybrid E/M phenotype(s) (Jia et al. 2017; Zhang et al. 2014). Importantly, these models have contributed to the experimental identification of hybrid E/M phenotypes at a single-cell level in multiple cancer types (Jolly et al. 2018b).

Further models built upon these frameworks have predicted what factors can stabilize a hybrid E/M phenotype (Fig. 1.3a) (Jolly et al. 2016; Hong et al. 2015). The role of 'phenotypic stability factors' (PSFs) has been experimentally validated; H1975 cells that stably maintained a hybrid E/M phenotype over 2 months *in vitro* switched to a mesenchymal state within a day upon knockdown of one of the PSFs – OVOL, GRHL2, or NUMB (Bocci et al. 2017; Jolly et al. 2016). Further, knockdown of one or more these PSFs has been observed to impair collective cell migration in developmental contexts as well such as mammary morphogenesis (Jolly et al. 2015).

Aside from these continuous approaches, Boolean models have also been developed for EMT networks (Font-Clos et al. 2018; Steinway et al. 2014, 2015). In these models, each node typically takes a value 0 (OFF) or 1 (ON) depending on the state of other nodes that activate or inhibit it. Thus, in this framework, a state is defined as epithelial epithelial when most markers or drivers of EMT are 'OFF', and that for MET are 'ON'. Conversely, it is defined as mesenchymal when most markers or drivers of MET are 'OFF', and that for EMT are 'ON'. The hybrid E/M state(s) tend to display a 'mixed' pattern, i.e. where many drivers or markers of EMT and MET are 'ON' or 'OFF' simultaneously. These models, typically developed for much larger networks, reinforce the earlier prediction that cells can stably maintain one or more hybrid E/M phenotype(s) (Fig. 1.3c). Moreover, these discrete models have led to identifying network motifs that can stabilize epithelial or mesenchymal states, complementing the analysis by continuous dynamic models – deduction of motifs to identify further PSFs (Jolly et al. 2016).

An alternative approach that has been implemented to investigate the dynamics of EMT is RACIPE (Random Circuit Perturbation) – a tool that generates an ensemble of mathematical

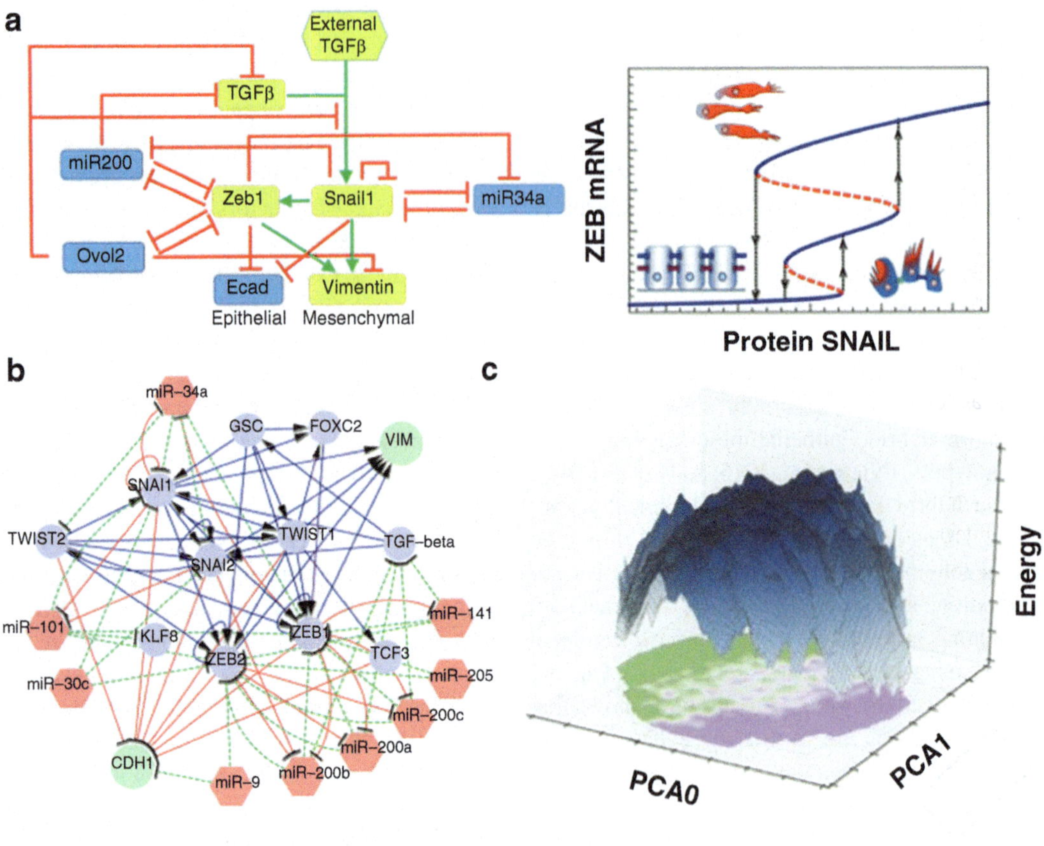

Fig. 1.3 Dynamics of epithelial-mesenchymal transition (EMT) (**a**) Left: A gene regulatory circuit for EMT including PSFs, as proposed by Hong et al. (2015) (https://journals.plos.org/ploscompbiol/article?id=10.1371/journal.pcbi.1004569) Right: a bifurcation diagram of ZEB mRNA as a function of EMT-TF SNAIL; it shows three stable phenotypes (i.e. continuous blue curves) corresponding to epithelial (low ZEB), hybrid E/M (intermediate ZEB) and mesenchymal (high ZEB) (http://www.pnas.org/content/110/45/18144). (**b**) An extended EMT regulatory circuit modeled using RACIPE (Huang et al. 2017) (https://journals.plos.org/ploscompbiol/article?id=10.1371/journal.pcbi.1005456) (**c**) The energy landscape of a large EMT regulatory circuit adapted from Boolean model shows two main minima (purple and green projections) corresponding to epithelial and mesenchymal phenotypes, respectively. Additionally, many local energy minima en route to EMT correspond to intermediate E/M states (Font-Clos et al. 2018) (http://www.pnas.org/content/115/23/5902.short). (Figure adapted from https://arxiv.org/pdf/1808.09113.pdf (Bocci et al. 2018))

models for a given network topology and infers the robust dynamical features of that topology (Huang et al. 2017). Each model in the ensemble is solved using continuous approaches (as ODEs) but the kinetic parameters are chosen randomly from a biologically relevant range, instead of from specific experiments regarding a given link/node. RACIPE analysis has been useful in strengthening the emerging notion that cells can stably adopt one or more hybrid E/M phenotype(s) (Fig. 1.3b), and in identifying the network links that can alter the likelihood of attaining the hybrid state(s) (Jolly et al. 2018a).

1.4 Conclusion

The biochemical signaling and mechanical models for collective cell migration have generated valuable insights into the subcellular and intercellular dynamics of this intricated process. Multiple modeling frameworks – both for mechanical and biochemical aspects – have been quite complementary in understanding how cells coordinate their migration in many contexts. To date, however, largely speaking, these models have been treated independently. While coupling them is not a straightforward task by any means, recent experiments have demonstrated many mechanochemical coupling (Vishwakarma et al. 2018; Riahi et al. 2015; Das et al. 2015) that can be considered as important initial steps in achieving this arduous task. Integrating biomechanical parameters such as cell-cell adhesion as a function of biochemical signaling components such as E-cadherin would require careful experimental calibration that can then be incorporated into integrated mechanical-biochemical model frameworks.

References

Albert PJ, Schwarz US (2016) Dynamics of cell ensembles on adhesive micropatterns: bridging the gap between single cell spreading and collective cell migration. PLOS Comp Biol 12(4):e1004863. https://doi.org/10.1371/journal.pcbi.1004863

Bi D, Lopez JH, Schwarz JM, Manning ML (2015) A density-independent rigidity transition in biological tissues. Nat Phys 11(12):1074–1079. https://doi.org/10.1038/nphys3471

Bi D, Yang X, Marchetti MC, Manning ML (2016) Motility-driven glass and jamming transitions in biological tissues. Phys Rev X 6(2):1–13. https://doi.org/10.1103/PhysRevX.6.021011

Bocci F, Jolly MK, Tripathi SC, Aguilar M, Hanash SM, Levine H, Onuchic JN (2017) Numb prevents a complete epithelial–mesenchymal transition by modulating notch signalling. J R Soc Interface 14(136):20170512. https://doi.org/10.1098/rsif.2017.0512

Bocci F, Levine H, Onuchic JN, Jolly MK, Jolly MK (2018) Deciphering the dynamics of epithelial-mesenchymal transition and cancer stem cells in tumor progression. https://arxiv.org/pdf/1808.09113. Accessed 30 Oct

Brugués A, Anon E, Conte V, Veldhuis JH, Gupta M, Colombelli J, Muñoz JJ, Brodland GW, Ladoux B, Trepat X (2014) Forces driving epithelial wound healing. Nat Phys 10(9):683–690. https://doi.org/10.1038/nphys3040

Camley BA, Rappel W-J (2017) Physical models of collective cell motility: from cell to tissue. J Phys D Appl Phys 50(11):113002. IOP Publishing. https://doi.org/10.1088/1361-6463/aa56fe

Camley BA, Zhang Y, Zhao Y, Li B, Ben-Jacob E, Levine H, Rappel W-J (2014) Polarity mechanisms such as contact inhibition of locomotion regulate persistent rotational motion of mammalian cells on micropatterns. Proc Natl Acad Sci 111(41):14770–14775. https://doi.org/10.1073/pnas.1414498111

Camley BA, Zimmermann J, Levine H, Rappel W-J (2016) Emergent collective chemotaxis without single-cell gradient sensing. Phys Rev Lett 098101 (March):1–6. https://doi.org/10.1103/PhysRevLett.116.098101

Cheung, KJ, Ewald AJ (2016) A collective route to metastasis: seeding by tumor cell clusters. Science (New York, NY) 352(6282):167–69. American Association for the Advancement of Science. https://doi.org/10.1126/science.aaf6546

Das T, Safferling K, Rausch S, Grabe N, Boehm H, Spatz JP (2015) A molecular mechanotransduction pathway regulates collective migration of epithelial cells. Nat Cell Biol 17(3):276–287. Nature Publishing Group. https://doi.org/10.1038/ncb3115

Font-Clos F, Zapperi S, La Porta CAM (2018) Topography of epithelial-mesenchymal plasticity. Proc Natl Acad Sci U S A 115(23):5902–5907. National Academy of Sciences. https://doi.org/10.1073/pnas.1722609115

Graner F, Glazier JA (1992) Simulation of biological cell sorting using a two-dimensional extended potts model. Phys Rev Lett 69(13):2013–2016. https://doi.org/10.1103/PhysRevLett.69.2013

Haeger A, Wolf K, Zegers MM, Friedl P (2015) Collective cell migration: guidance principles and hierarchies. Trends Cell Biol 25(9):556–566. https://doi.org/10.1016/j.tcb.2015.06.003

Honda H (1978) Description of cellular patterns by dirichlet domains: the two-dimensional case. J Theor Biol 72(3):523–543. https://doi.org/10.1016/0022-5193(78)90315-6

Hong T, Watanabe K, Ta CH, Villarreal-Ponce A, Nie Q, Dai X (2015) An Ovol2-Zeb1 mutual inhibitory circuit governs bidirectional and multi-step transition between epithelial and mesenchymal states. Edited by Michael PH. Stumpf. PLOS Comp Biol 11(11):e1004569. Public Library of Science. https://doi.org/10.1371/journal.pcbi.1004569

Huang, B, Lu M, Jia D, Ben-Jacob E, Levine H, Onuchic JN (2017) Interrogating the topological robustness of gene regulatory circuits by randomization. Edited by Chao Tang. PLOS Comp Biol 13(3):e1005456. Public Library of Science. https://doi.org/10.1371/journal.pcbi.1005456

Jia D, Jolly MK, Tripathi SC, Den Hollander P, Huang B, Lu M, Celiktas M et al (2017) Distinguishing mechanisms underlying EMT tristability. Cancer Converg 1(1):2. Springer. https://doi.org/10.1186/s41236-017-0005-8

Jolly MK, Boareto M, Huang B, Jia D, Mingyang L, Ben-Jacob E, Onuchic JN, Levine H (2015) Implications of the hybrid epithelial/Mesenchymal phenotype in metastasis. Front Oncol 5(July):155. Frontiers. https://doi.org/10.3389/fonc.2015.00155

Jolly MK, Tripathi SC, Jia D, Mooney SM, Celiktas M, Hanash SM, Mani SA, Pienta KJ, Ben-Jacob E, Levine H (2016) Stability of the hybrid epithelial/Mesenchymal phenotype. Oncotarget 7(19):27067–27084. https://doi.org/10.18632/oncotarget.8166

Jolly MK, Preca B-T, Tripathi SC, Jia D, George JT, Hanash SM, Brabletz T, Stemmler MP, Maurer J, Levine H (2018a) Interconnected feedback loops among ESRP1, HAS2, and CD44 Regulate epithelial-mesenchymal plasticity in cancer. APL Bioeng 2(3):031908. AIP Publishing LLC. https://doi.org/10.1063/1.5024874

Jolly MK, Somarelli JA, Sheth M, Biddle A, Tripathi SC, Armstrong AJ, Hanash SM, Bapat SA, Rangarajan A, Levine H (2018b) Hybrid epithelial/Mesenchymal phenotypes promote metastasis and therapy resistance across carcinomas. Pharmacol Ther. https://doi.org/10.1016/j.pharmthera.2018.09.007

Kalluri, R, Weinberg RA (2009) The basics of epithelial-mesenchymal transition. J Clin Invest 119(6):1420–28. American Society for Clinical Investigation. https://doi.org/10.1172/JCI39104

Koride S, Loza AJ, Sun SX, Koride S, Loza AJ, Sun SX (2018) Epithelial vertex models with active biochemical regulation of contractility can explain organized collective cell motility epithelial vertex models with active biochemical regulation of contractility can explain organized collective cell motility. APL Bioeng:031906. https://doi.org/10.1063/1.5023410

Löber J, Ziebert F, Aranson IS (2015) Collisions of deformable cells lead to collective migration. Sci Rep 5:1–7. https://doi.org/10.1038/srep09172

Lu M, Jolly MK, Levine H, Onuchic JN, Ben-Jacob E (2013) MicroRNA-based regulation of epithelial-hybrid-mesenchymal fate determination. Proc Natl Acad Sci U S A 110(45):18144–18149. National Academy of Sciences. https://doi.org/10.1073/pnas.1318192110

Martin P, Shaw TJ, Martin P, Lecuit T, Lenne P-F, Supplementary Fig, Trepat X et al (2015) Gap geometry dictates epithelial closure efficiency. Nat Commun 6(1):7683. The Author(s). https://doi.org/10.1038/nrm2222

Micalizzi DS, Farabaugh SM, Ford HL (2010) Epithelial-Mesenchymal transition in cancer: parallels between normal development and tumor progression. J Mammary Gland Biol Neoplasia 15(2):117–134. https://doi.org/10.1007/s10911-010-9178-9

Nieto M, Angela R, Huang Y-J, Jackson RA, Thiery JP (2016) EMT: 2016. Cell 166(1):21–45. Cell Press. https://doi.org/10.1016/J.CELL.2016.06.028

Petitjean L, Reffay M, Grasland-Mongrain E, Poujade M, Ladoux B, Buguin A, Silberzan P (2010) Velocity fields in a collectively migrating epithelium. Biophys J 98(9):1790–1800. Biophysical Society. https://doi.org/10.1016/j.bpj.2010.01.030

Poujade M, Grasland-Mongrain E, Hertzog A, Jouanneau J, Chavrier P, Ladoux B, Buguin A, Silberzan P (2007) Collective migration of an epithelial monolayer in response to a model wound. Proc Natl Acad Sci 104(41):15988–15993. https://doi.org/10.1073/pnas.0705062104

Prost J, Jülicher F, Joanny J-F (2015) Active gel physics. Nat Phys 11(2):111–117. Nature Publishing Group. https://doi.org/10.1038/nphys3224

Reffay M, Petitjean L, Coscoy S, Grasland-Mongrain E, Amblard F, Buguin A, Silberzan P (2011) Orientation and polarity in collectively migrating cell structures: statics and dynamics. Biophys J 100(11):2566–2575. Biophysical Society. https://doi.org/10.1016/j.bpj.2011.04.047

Reffay M, Parrini MC, Cochet-Escartin O, Ladoux B, Buguin A, Coscoy S, Amblard F, Camonis J, Silberzan P (2014) Interplay of RhoA and mechanical forces in collective cell migration driven by leader cells. Nat Cell Biol 16(3):217–223. https://doi.org/10.1038/ncb2917

Riahi R, Sun J, Wang S, Long M, Zhang DD, Wong PK (2015) Notch1–Dll4 signalling and mechanical force regulate leader cell formation during collective cell migration. Nat Commun 6(1):6556. Nature Publishing Group. https://doi.org/10.1038/ncomms7556

Sandersius SA, Newman TJ (2008) Modeling cell rheology with the subcellular element model. Phys Biol 5(1): 015002. IOP Publishing. https://doi.org/10.1088/1478-3975/5/1/015002

Sandersius SA, Weijer CJ, Newman TJ (2011) Emergent cell and tissue dynamics from subcellular modeling of active biomechanical processes. Phys Biol 8(4): 045007. IOP Publishing. https://doi.org/10.1088/1478-3975/8/4/045007

Scarpa E, Mayor R (2016) Collective cell migration in development. JCell Biol 212(2):143–155. The Rockefeller University Press. https://doi.org/10.1083/jcb.201508047

Segerer FJ, Thüroff F, Alberola AP, Frey E, Rädler JO (2015) Emergence and persistence of collective cell migration on small circular micropatterns. Phys Rev Lett 114(22):228102. American Physical Society. https://doi.org/10.1103/PhysRevLett.114.228102

Sepúlveda N, Petitjean L, Cochet O, Grasland-Mongrain E, Silberzan P, Hakim V (2013) Collective cell motion in an epithelial sheet can be quantitatively described by a stochastic interacting particle model. PLoS Comput Biol 9(3). https://doi.org/10.1371/journal.pcbi.1002944

Shao D, Rappel W-J, Levine H (2010) Computational model for cell morphodynamics. Phys Rev Lett 105(10):108104. https://doi.org/10.1103/PhysRevLett.105.108104

Shao D, Levine H, Rappel W-J (2012) Coupling actin flow, adhesion, and morphology in a computational cell motility model. Proc Natl Acad Sci 109(18):6851–6856. https://doi.org/10.1073/pnas.1203252109

Steinway SN, Zañudo JGT, Ding W, Rountree CB, Feith DJ, Loughran TP, Albert R, Albert R (2014) Network modeling of TGFβ signaling in hepatocellular carcinoma epithelial-to-mesenchymal transition reveals joint sonic hedgehog and wnt pathway activation. Cancer Res 74(21):5963–5977. NIH Public Access. https://doi.org/10.1158/0008-5472.CAN-14-0225

Steinway SN, Zañudo JGT, Michel PJ, Feith DJ, Loughran TP, Albert R (2015) Combinatorial interventions Inhibit TGFβ-Driven epithelial-to-mesenchymal transition and support hybrid cellular phenotypes. Npj Sys Biol Appl 1(1):15014. Nature Publishing Group. https://doi.org/10.1038/npjsba.2015.14

Szabó A, Ünnep R, Méhes E, Twal WO, Argraves WS, Cao Y, Czirók A (2010) Collective cell motion in endothelial monolayers. Phys Biol 7(4): 046007. IOP Publishing. https://doi.org/10.1088/1478-3975/7/4/046007

Tarle V, Ravasio A, Hakim V, Gov NS (2015) Modeling the finger instability in an expanding cell monolayer. Integr Biol (United Kingdom) 7(10):1218–1227. Royal Society of Chemistry. https://doi.org/10.1039/c5ib00092k

Tarle V, Gauquelin E, Vedula SRK, D'Alessandro J, Lim CT, Ladoux B, Gov NS (2017) Modeling collective cell migration in geometric confinement. Phys Biol 14(3):035001. IOP Publishing. https://doi.org/10.1088/1478-3975/aa6591

te Boekhorst V, Preziosi L, Friedl P (2016) Plasticity of cell migration in vivo and in silico. Annu Rev Cell Dev Biol 32(1):491–526. https://doi.org/10.1146/annurev-cellbio-111315-125201

Vedula SRK, Peyret G, Cheddadi I, Chen T, Brugués A, Hirata H, Lopez-Menendez H et al (2015) Mechanics of epithelial closure over non-adherent environments. Nat Commun 6(January):6111. Nature Publishing Group. https://doi.org/10.1038/ncomms7111

Vicsek T, Czirók A, Ben-Jacob E, Cohen I, Shochet O (1995) Novel type of phase transition in a system of self-driven particles. Phys Rev Lett 75(6):1226–1229. https://doi.org/10.1103/PhysRevLett.75.1226

Vishwakarma M, Di Russo J, Probst D, Schwarz US, Das T, Spatz JP (2018) Mechanical interactions among followers determine the emergence of leaders in migrating epithelial cell collectives. Nat Commun 9(1):3469. Nature Publishing Group. https://doi.org/10.1038/s41467-018-05927-6

Wigner E, Seitz F (1933) On the constitution of metallic sodium. Phys Rev 43(10):804–810. https://doi.org/10.1103/PhysRev.43.804

Yang Y, Levine H (2018) Role of the supracellular actomyosin cable during epithelial wound healing. Soft Matter 14(23):4866–4873. https://doi.org/10.1039/C7SM02521A

Zhang J, Tian X-J, Zhang H, Teng Y, Li R, Bai F, Elankumaran S, Xing J (2014) TGF-β-induced epithelial-to-mesenchymal transition proceeds through stepwise activation of multiple feedback loops. Sci Signal 7(345):ra91. American Association for the Advancement of Science. https://doi.org/10.1126/scisignal.2005304

Ziebert F, Swaminathan S, Aranson IS (2012) Model for self-polarization and motility of keratocyte fragments. J R Soc Interface 9(70):1084–92. The Royal Society. https://doi.org/10.1098/rsif.2011.0433

Zimmermann J, Camley BA, Rappel W-j, Levine H (2016) Contact inhibition of locomotion determines cell – cell and cell – substrate forces in tissues. Proc Natl Acad Sci 113(10). https://doi.org/10.1073/pnas.1522330113

Contour Models of Cellular Adhesion

2

Luca Giomi

Abstract
The development of traction-force microscopy, in the past two decades, has created the unprecedented opportunity of performing direct mechanical measurements on living cells as they adhere or crawl on uniform or micro-patterned substrates. Simultaneously, this has created the demand for a theoretical framework able to decipher the experimental observations, shed light on the complex biomechanical processes that govern the interaction between the cell and the extracellular matrix and offer testable predictions. Contour models of cellular adhesion, represent one of the simplest and yet most insightful approach in this problem. Rooted in the paradigm of active matter, these models allow to explicitly determine the shape of the cell edge and calculate the traction forces experienced by the substrate, starting from the internal and peripheral contractile stresses as well as the passive restoring forces and bending moments arising within the actin cortex and the plasma membrane. In this chapter I provide a general overview of contour models of cellular adhesion and review the specific cases of cells equipped with isotropic and anisotropic actin cytoskeleton as well as the role of bending elasticity.

Keywords
Cell mechanics · Cell adhesion · Contour models

In this chapter we review one of the simplest and yet most insightful theoretical approach for modeling the mechanics of cell-substrate interaction, based on two-dimensional contour models. Introduced by Bar-Ziv et al. to account for the pearling instability in fibroblasts (Bar-Ziv et al. 1999) and later systematically developed by Bischofs et al. (2008, 2009) in the context of traction force microscopy, this approach consists of modeling cells adhering on uniform or micro-patterned substrates as two-dimensional active gels subject to internal and external forces. This approach allows an explicit calculation of both the shape of the cell and the traction forces exerted by the cell on the substrate, under the assumption that the time scale required for the equilibration of the internal forces is much shorter than the typical time required by the cell to move across the substrate (i.e. minutes). The seemingly intractable problem of predicting the shape of a living cell is then brought into the realm of classical continuum mechanics, with the task of modeling the

L. Giomi (✉)
Instituut-Lorentz, Universiteit Leiden, Leiden, The Netherlands
e-mail: giomi@lorentz.leidenuniv.nl

C. A. M. La Porta, S. Zapperi (eds.), *Cell Migrations: Causes and Functions*, Advances in Experimental Medicine and Biology 1146,
https://doi.org/10.1007/978-3-030-17593-1_2

internal passive and active forces representing the major technical challenge. Starting from the original work by Bar-Ziv et al. (1999), much progress has been made on incorporating into this simple picture aspects of the mechanical complexity of eukaryotic cells, including the bending elasticity of the plasma membrane (Banerjee and Giomi 2013) and the anisotropy of the actin cytoskeleton (Pomp et al. 2018). Whether the final product of this reductionist effort is still far from providing a theoretical framework whose predictive power is comparable with the most recent computational approaches (Sabass et al. 2008), it nevertheless represents an indispensable conceptual step toward a satisfactory understanding of the physical properties of the cell.

From simple prokaryotes to the more complex eukaryotes, living cells are capable of astonishing mechanical functionalities (Janmey and McCulloch 2007). They can repair wounded tissues by locally contracting the extracellular matrix (Midwood et al. 2004), move in a fluid or on a substrate (Barry and Bretscher 2010), and generate enough force to split themselves in two while remaining alive (Tanimoto and Sano 2012). Conversely, cell behavior and fate crucially depend on mechanical cues from outside the cell (Discher et al. 2005; Geiger et al. 2009; Jülicher et al. 2007; Mendez and Janmey 2012). Examples include rigidity-dependent stem cell differentiation (Engler et al. 2006; Trappmann et al. 2012), protein expression regulated by internal stresses (Sawada et al. 2006), mechanical cell-cell communication (Reinhart-King et al. 2008) and durotaxis (Lo et al. 2000; Sochol et al. 2011). In all these bio-mechanical processes, cells rely on their shape to gauge the mechanical properties of their microenvironment (Ghibaudo et al. 2009) and direct the traction forces exerted on their surrounding (Schwarz and Safran 2013).

Immediately after coming into contact with such a surface, many animal cells spread and develop transmembrane adhesion receptors. This induces the actin cytoskeleton to reorganize into cross-linked networks and bundles (Burridge and Chrzanowska-Wodnicka 1996; Burridge and Wittchen 2013), whereas adhesion becomes limited to a number of sites, distributed mainly along the cell contour (i.e. focal adhesions Burridge and Chrzanowska-Wodnicka 1996). At this stage, cells are essentially flat and assume a typical shape characterized by arcs which span between the sites of adhesion (Fig. 2.1a), while forces are mainly contractile (Schwarz and Safran 2013). On timescales much shorter than those required to change its shape, the cell can be considered in mechanical equilibrium at any point of its interface. Upon treating the cell as a two-dimensional continuum separated by the surrounding environment by a one-dimensional interface, such an equilibrium condition translates into the following force balance relation at the cell cortex:

$$\boldsymbol{F}' + (\hat{\boldsymbol{\Sigma}}_{\rm out} - \hat{\boldsymbol{\Sigma}}_{\rm in}) \cdot \boldsymbol{N} + \boldsymbol{f}_{\rm ext} = \boldsymbol{0}, \tag{2.1}$$

where $\boldsymbol{F}$ is the stress resultant along the cell cortex, the prime indicates differentiation with respect to the arc-length of the cell edge (i.e. $\boldsymbol{F}' = {\rm d}\boldsymbol{F}/{\rm d}s$), $\hat{\boldsymbol{\Sigma}}_{\rm out}$ and $\hat{\boldsymbol{\Sigma}}_{\rm in}$ are the stress tensors outside and inside the cell and $\boldsymbol{f}_{\rm ext}$ is an external force per unit length, possibly resulting from a deformation of the substrate. The cell contour is parametrized as a plane curve spanned by the arc-length s and oriented along the inward pointing normal vector $\boldsymbol{N}$ (Fig. 2.1b).

Equation (2.1) is the starting point of all contour models of cellular adhesion as well as the core of this chapter. Before embarking on analyzing specific cases it is worth to stress that Eq. (2.1) is rooted into the following two fundamental assumptions: (1) cellular motion occurs *adiabatically*, because to the large separation of time scales associated with force relaxation and cell migration; (2) as a consequence of the flat morphology of adherent cells, one can ignore out-of-plane forces and reduce the dimension of the problem from three to two.

In the remaining of this chapter, we will see how different bio-mechanical scenarios result in different choices of the forces and stresses involved in Eq. (2.1) and how these affect the shape of the cells and the traction forces experienced by the substrate. The chapter is organized as follows: in Sect. 2.1 we review some simple concept on plane curves and we fix a notation; in Sect. 2.2

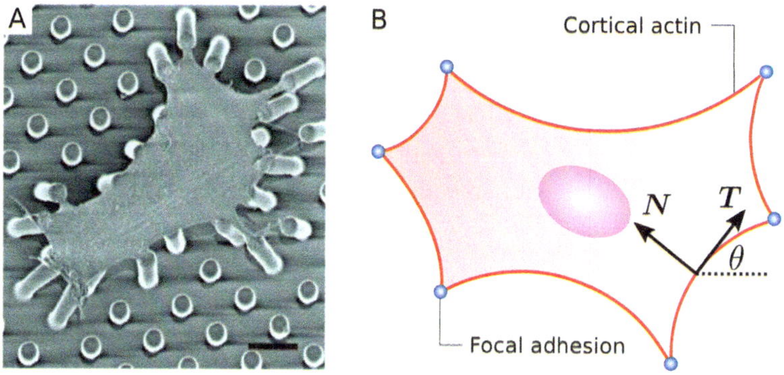

Fig. 2.1 (**a**) Scanning electron micrograph of a smooth muscle cell attached to an array of posts coated at the tip with fibronectin. Scale bar 10 μm. Adapted from Tan et al. (2003). (**b**) Schematic representation of a contour model of adherent cells. The cell edge is parameterized as closed plane curve comprising concave arcs connecting pairs of adhesion points (blue dots). The geometry of the arcs can be entirely described via the two-dimensional Frenet-Serret frame consisting of the tangent vector $\boldsymbol{T} = (\cos\theta, \sin\theta)$ and the normal vector $\boldsymbol{N} = (-\sin\theta, \cos\theta)$, with θ the turning angle

we introduce Bischofs' et al. simple tension model for isotropic contractile cells; in Sect. 2.3 we discuss the case of anisotropic cells in the framework of the anisotropic tension model; in Sect. 2.4 we explore the effect of bending elasticity; Sect. 2.5 is devoted to conclusive remarks.

2.1 Mathematical Preliminaries and Notation

Whereas different in the mechanical details, all contour models of cellular adhesion revolve around modeling cells as flat two-dimensional objects whose edge can be mathematically described as a closed, but not necessarily smooth, plane curve. The geometry of plane curves can be entirely described via a two-dimensional version of the Frenet-Serret frame, consisting of the tangent vector $\boldsymbol{T}$ and the normal vector $\boldsymbol{N}$ (Fig. 2.1b). Letting $\boldsymbol{r} = \boldsymbol{r}(s)$ the position vector of a curve in $\mathbb{R}^2$, parametrized via the arc-length s, these are defined by the following identities:

$$\boldsymbol{r}' = \boldsymbol{T}, \qquad \boldsymbol{T}' = \kappa \boldsymbol{N}, \qquad \boldsymbol{N}' = -\kappa \boldsymbol{T}, \tag{2.2}$$

where $\kappa = \kappa(s)$ is the local curvature of the curve. Equation (2.2) describe how the orthonormal frame $\{\boldsymbol{T}, \boldsymbol{N}\}$ rotates on the plane as we move along the curve. Once κ is assigned, in the form of a differentiable function, the fundamental theorem of plane curves guarantees that the corresponding curve is uniquely determined, up to a rigid motion (do Carmo 1976). Furthermore, the tangent vector can be parametrized through a single scalar function $\theta = \theta(s)$, representing the turning angle of $\boldsymbol{T}$. In a standard Cartesian frame: $\boldsymbol{T} = (\cos\theta, \sin\theta)$, then, using Eq. (2.2), one finds $\kappa = \mathrm{d}\theta/\mathrm{d}s$ and:

$$\boldsymbol{r}(s) = \boldsymbol{r}(0) + \int_0^s \mathrm{d}s' \left[\cos\theta(s')\,\hat{\boldsymbol{x}} + \sin\theta(s')\,\hat{\boldsymbol{y}}\right], \tag{2.3}$$

with $\theta(s) = \theta(0) + \int_0^s \mathrm{d}s\,\kappa(s')$. Expressing the curvature of a plane curve as the derivative of the turning angle θ, allows us to unambiguously identify the sign of κ (unlike in three dimensions). Thus $\kappa > 0$ ($\kappa < 0$) implies that the turning angle increases (decreases) as we move along the curve, whereas reversing the orientation of the curve changes the sign of κ. If a closed plane curve is oriented counterclockwise and the turning angle is measured, as usual, with respect to the x−axis, then $\kappa > 0$ ($\kappa < 0$) will then corresponds to points where the curve is convex (concave).

Closed curves, have a number of interesting global properties that serve as important calculation tools in the later sections. The four vertex theorem (DeTurck et al. 2007), is one of the earliest results in global differential geometry and states that the curvature of a simple, smooth, closed curve on a plane has at least four vertices: i.e. four extrema where $\mathrm{d}\kappa/\mathrm{d}s = 0$ (specifically two maxima and two minima). Another fundamental property of closed plane curves is expressed by the theorem of turning tangents (do Carmo 1976; Gray 1997). Namely:

$$\oint \mathrm{d}s\, \kappa = 2\pi m. \tag{2.4}$$

The integer m is called the rotation index of the curve and measures how many times the tangent vector turns with respect to a fixed direction (Gray 1997). Simple closed curves have thus $m = 1$, whereas a curve that loops twice around its center (thus self-intersects once before closing) has $m = 2$. If a simple closed curve has kinks (i.e. singular points where the tangent vector switches discontinuously between two orientations), these will affect the total curvature as follows:

$$\oint \mathrm{d}s\, \kappa + \sum_i \vartheta_i = 2\pi m. \tag{2.5}$$

where ϑ_i is the external angle at each kink and the summation runs over all the kinks. In the case of a convex polygon, for instance, $\kappa = 0$ and Eq. (2.5) asserts that the sum of the external angles is equal to 2π.

Finally, by virtue of the divergence theorem in two dimensions, the area enclosed by a plane curve can be expressed as a contour integral as follows:

$$\int \mathrm{d}A = -\frac{1}{2}\oint \mathrm{d}s\, \boldsymbol{N}\cdot\boldsymbol{r}, \tag{2.6}$$

where the minus sign results from the convention of choosing the normal vector directed toward the interior of the curve.

2.2 Simple Tension Model

At time scales when the cell is fully spread, the forces generated by the actomyosin cytoskeleton are mainly contractile and this gives rise to an *effective tension* that is transmitted to the substrate thorough the focal adhesions (Bar-Ziv et al. 1999). The actin cortex localized at the cell periphery naturally resists to this inward contraction by generating a competing contractile force, which, in turn, balances the bulk contractility leading to an equilibrium configuration. The overall effect of actomyosin contractility and adhesion on the shape of the cell was investigated by Bischofs et al. in (2008, 2009) by mean of a simple and yet very rich mechanical model known as *simple tension model.*

If the contractile forces acting in the interior of the cell are isotropic, the difference $\hat{\boldsymbol{\Sigma}}_{\rm out} - \hat{\boldsymbol{\Sigma}}_{\rm in}$ between the stresses across the cell edge can be reasonably approximated as an active pressure, namely:

$$\hat{\boldsymbol{\Sigma}}_{\rm out} - \hat{\boldsymbol{\Sigma}}_{\rm in} = \sigma \hat{\boldsymbol{I}}, \tag{2.7}$$

where $\hat{\boldsymbol{I}}$ is the two-dimensional identity matrix. Similarly, the stress resultant produced by the contraction of peripheral actin can be expressed as $\boldsymbol{F} = \lambda \boldsymbol{T}$, where $\lambda = \lambda(s)$ is an effective interfacial tension. In the absence of external forces, Eqs. (2.1) and (2.2) yield:

$$\lambda' \boldsymbol{T} + (\lambda + \kappa\sigma)\boldsymbol{N} = \boldsymbol{0}, \tag{2.8}$$

along each individual cellular arc connecting two consecutive adhesion points. Because $\boldsymbol{T}$ and $\boldsymbol{N}$ are orthogonal, both terms must vanish. Thus, mechanical equilibrium requires $\lambda = \text{const}$ and:

$$\kappa = -\frac{\sigma}{\lambda}, \tag{2.9}$$

where the negative sign reflects that the cell is everywhere concave, with exception for a discrete number of adhesion points. The same result could have also been obtained from a minimization of an effective energy functional of the from:

$$E[\boldsymbol{r}] = \lambda \oint \mathrm{d}s + \sigma \int \mathrm{d}A. \tag{2.10}$$

Indeed, upon performing a small normal variation of the curve, $\boldsymbol{r} \to \boldsymbol{r} + \epsilon \boldsymbol{N}$, standard manipulations of Eq. (2.10) yield:

$$\delta E = E[\boldsymbol{r} + \epsilon \boldsymbol{N}] - E[\boldsymbol{r}] = -\oint \mathrm{d}s\,(\sigma + \lambda\kappa)\epsilon. \tag{2.11}$$

Thus Eq. (2.9) identifies a minimizer of the energy functional Eq. (2.10). Equation (2.1) is, however, more generic and, as we will see in Sect. 2.3, allows to account for active stresses that could not be constructed from variational principles.

The quantities σ and λ are material parameters that embody the biomechanical activity of myosin motors in the actin cytoskeleton. Treating cells with pharmacological drugs able to disrupt the cytoskeleton allow some degree of manipulation of these quantities. Cells treated with Y-27632, a general inhibitor of the Rho-kinase pathway, and blebbistatin, a specific inhibitor of nonmuscle myosin II, have been reported to invaginate more than untreated cells, suggesting a strong reduction of λ accompanied by the presence of a residual σ (Bischofs et al. 2008). The competition between bulk and peripheral contractility along the cell boundary results, therefore, in the formation of arcs of constant curvature, through a mechanism analogous to the Young-Laplace law for fluid interfaces. The shape of the cell boundary is then approximated by a sequence of circular arcs, whose radius $R = 1/\kappa$ might or might not be uniform across the cell, depending on how the cortical tension λ varies from arc to arc (Fig. 2.2).

Although deliberately simple-minded, the picture outline so far is still sufficiently complex to account for some of the peculiar mechanical properties that are unique to living materials, such as the ability of regulating the forces exerted on a substrate depending on its stiffness. To

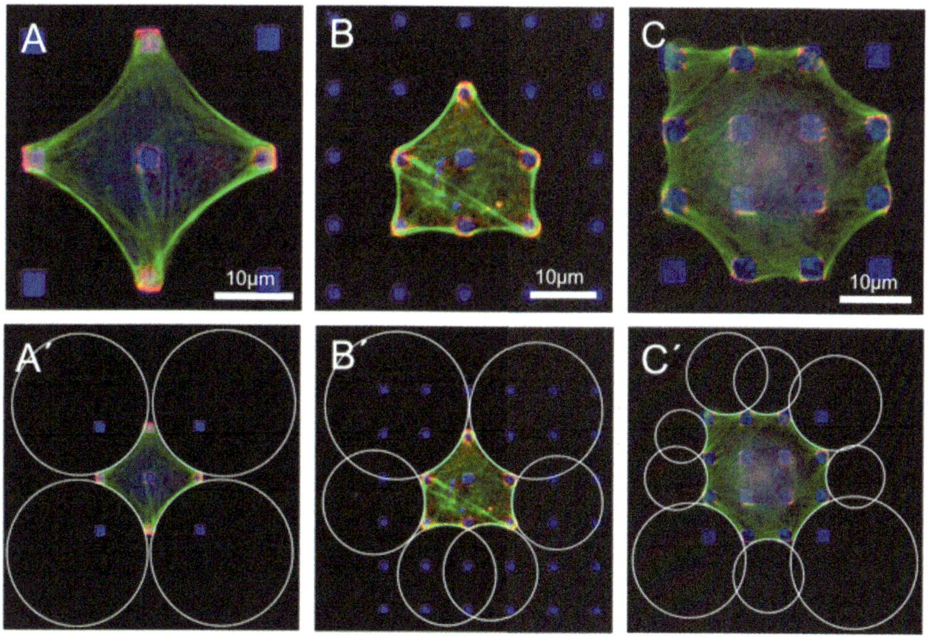

Fig. 2.2 Cell shape on micropatterned substrates. (**a–c**) Arc-like contours composed of actin fibers characterize the shape of BRL (**a** and **b**) and B16 cells (**c**) cultured on substrates of micropatterned fibronectin dots. Cultures were labeled for actin (green), paxillin (red), and fibronectin (blue). Scale bars 10 μm. (**a′c′**) For all cases, arc-like contours fit well to circles determined by custom-made software. (**b** and **c**) The circles spanning diagonal distances show larger radii than the circles spanning the shorter distances between neighboring adhesions. (Reproduced from Bischofs et al. 2008)

appreciate this, let us consider again Eq. (2.1). In between two adhesion points force balance results in the invagination of the cellular arcs as demanded by Eq. (2.9). At the adhesion points, on the other hand, both contractile forces exerted by the bulk and peripheral cytoskeleton are counterbalanced by the substrate, so that $\boldsymbol{f}_{\rm ext}(\boldsymbol{r}) = -\sum_i \boldsymbol{F}_{\rm traction}(\boldsymbol{r})\delta(\boldsymbol{r}-\boldsymbol{r}_i)$, where the sum runs over all adhesion points and the negative sign reflects that the force exerted by the cell on the substrate is equal and opposite to the force exerted by the substrate on the cell. Thus, from Eq. (2.1), we have:

$$\sum_i \boldsymbol{F}_{\rm traction}(\boldsymbol{r})\delta(\boldsymbol{r}-\boldsymbol{r}_i) = (\sigma+\lambda\kappa)\boldsymbol{N}. \quad (2.12)$$

Note that both sides of this equation diverge at the adhesion point as $\kappa \to \infty$ at a kink. In order to calculate the traction force exerted by a specific focal adhesion, one can approximate the kink as a circle of radius ϵ, hence curvatures $\kappa = 1/\epsilon$ (Fig. 2.3a). Then, taking $\boldsymbol{N} = (\cos\phi, \sin\phi)$, with $-\varphi \le \phi \le \varphi$, integrating Eq. (2.12) along the circle and taking the limit of $\epsilon \to 0$ yields (Bischofs et al. 2009):

$$\begin{aligned}\boldsymbol{F}_{\rm traction} &= \lim_{\epsilon\to 0}\int_{-\varphi}^{\varphi} {\rm d}\phi\,\epsilon\left(\sigma+\frac{\lambda}{\epsilon}\right)\boldsymbol{N}\\ &= \boldsymbol{F}(\varphi)+\boldsymbol{F}(-\varphi) = 2\lambda\cos\frac{\vartheta}{2}\,\hat{\boldsymbol{x}},\end{aligned} \quad (2.13)$$

where $\vartheta = \pi - 2\varphi$ is the opening angle of the kink (Fig. 2.3a).

This simple considerations illustrate the essence of adaptive mechanical response in cells. According to Eq. (2.13), the more acute is the kink (the smaller is ϑ) the larger is the force exerted by the cell. The maximum traction $\boldsymbol{T} = 2\lambda\hat{\boldsymbol{x}}$ is attained when $\vartheta = 0$, thus the kink reduces to a cusp and all the tension exerted by the peripheral actin is employed to deform the substrate. Now, the opening angle ϑ at a given adhesion point, is not a free parameter, but is set by the curvature and the length of the cellular arcs that are directly connected to it. To see this let us consider a simple triangular cell whose adhesion points are located at a distance $d = 2R\cos\psi$, with $R = \lambda/\sigma$ and ψ as illustrated in Fig. 2.3b, from one another and let $\bar{\psi}$ be the internal angle of the triangle identified by the convex hull of the cell at a specific adhesion point (Fig. 2.3b). The opening angle is then $\vartheta = \Theta - 2\bar{\psi}$, with $\psi + \bar{\psi} = \pi/2$. Therefore:

$$\cos\frac{\vartheta}{2} = \cos\frac{\Theta}{2}\sin\psi + \sin\frac{\Theta}{2}\cos\psi. \quad (2.14)$$

Finally, taking $\cos\psi = d\sigma/(2\lambda)$, one can express the traction force as:

$$\boldsymbol{F}_{\rm traction} = 2\lambda\left[\left(\frac{d\sigma}{2\lambda}\right)\sin\frac{\Theta}{2} + \sqrt{1-\left(\frac{d\sigma}{2\lambda}\right)^2}\cos\frac{\Theta}{2}\right]\hat{\boldsymbol{x}}. \quad (2.15)$$

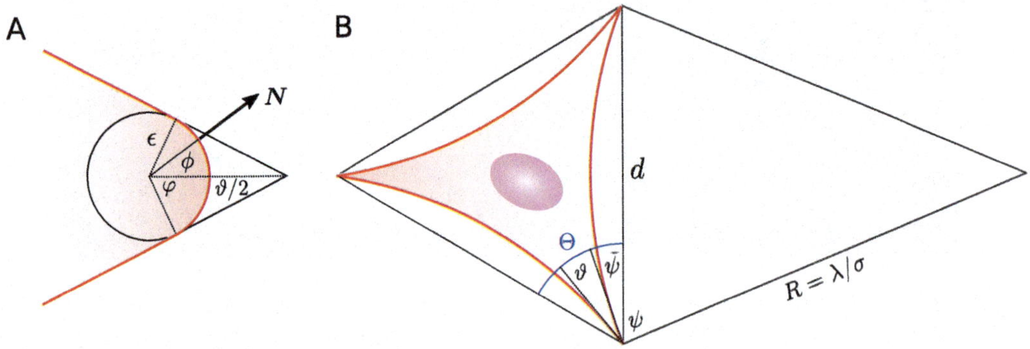

Fig. 2.3 (**a**) To calculate the traction force exerted by the cell at a point of adhesion, it is convenient to approximate a kink with a circular arc of radius ϵ and take the limit of $\epsilon \to 0$. This procedure yields Eq. (2.13). (**b**) Illustration of the calculation summarized by Eqs. (2.14) and (2.15)

Thus, for fixed σ and λ values, the more distant are the adhesion points, the more acute is the opening angle and the larger is the force exerted by the cell. Furthermore, if the substrate is compliant, the distance d between focal adhesion depends on how much it deforms under the effect of traction forces. In particular, the softer the substrate the closer are the adhesion points, the weaker is the force exerted by the cell. Viceversa, on stiffer substrates d will be larger and the cell is expected to exert more force. These predictions, which are verified in experiments with fibroblasts and endothelial cells plated on continuous substrates of various rigidity (Lo et al. 2000; Yeung et al. 2005), provide a simple and yet insightful example of how in the interplay between cellular geometry and the active contraction provided by the actin cytoskeleton, can lead to an adaptive mechanical behavior even in the absence of biochemical regulation.

2.3 Anisotropic Tension Model

Many cells, including the fibroblastoids (GDβ1, GDβ3) and epithelioids (GEβ1, GEβ3) displayed in Fig. 2.4a (Danen et al. 2002; Pomp et al. 2018), develop directed forces by virtue of the strong anisotropic cytoskeleton originating from the actin stress fibers (Burridge and Wittchen 2013; Pellegrin and Mellor 2007). This scenario is, evidently, beyond the scope of the simple tension model reviewed in Sect. 2.2. In these cells, the longer arcs appear indeed prominently non-circular, as indicated by the fact that their curvature smoothly varies along the arc by a factor two (Fig. 2.4b). On the other hand, the average radius of curvature of the cellular arcs appears significantly correlated with the orientation of the stress fibers. In particular, the radius of curvature decreases as the stress fibers become more perpendicular to the cell cortex (Fig. 2.4c).

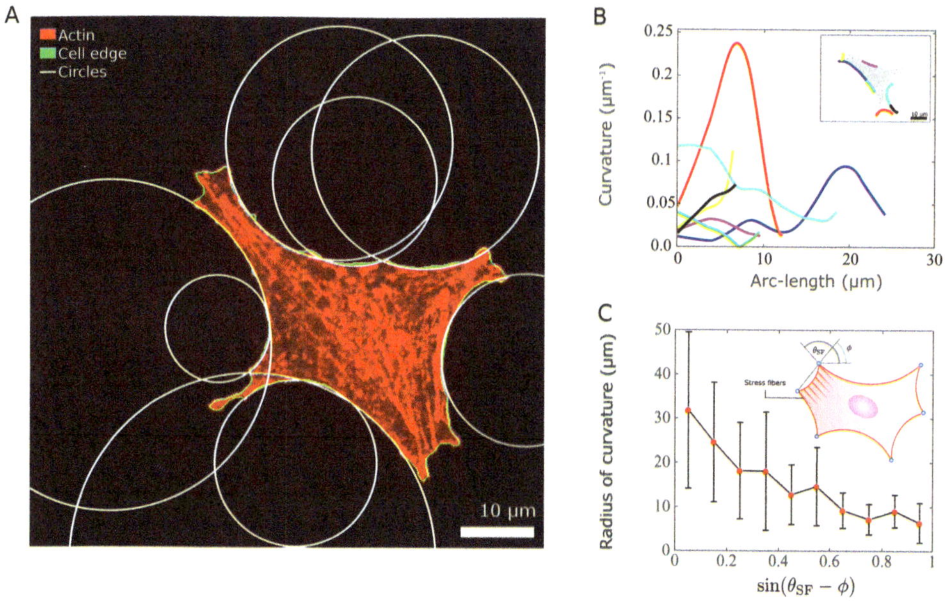

Fig. 2.4 Relation between stress fibers and curvature of the cell edge. (**a**) A cell with an anisotropic actin cytoskeleton (epithelioid GEβ3) with circles (white) fitted to its edges (green). The end-points of the arcs are identified based on the forces exerted on the pillars. The actin cytoskeleton is visualized with TRITC-Phalloidin (red). Scalebar is 10 μm. (**b**) Curvature versus arc-length for a specific cell (inset). Longer arcs, whose length is much larger than their average radius of curvature (i.e. $L \gg 1/|\kappa|$), are evidently non-circular as indicated by the fact that their curvature smoothly varies by a factor two along the arc. (**c**) Arc radius as a function of the sine of the angle $\theta_{SF} - \phi$, between the local orientation of the stress fibers and that of the distance between the adhesion points (data show the mean ± standard deviation). (Adapted from Pomp et al. 2018)

This correlation is intuitive as the bulk contractile stress focusses in the direction of the stress fibers.

The anisotropy of the actin cytoskeleton can be incorporated into the contour models by modelling the stress fibers as contractile force-dipoles. As it is known from the literature on active fluids (Pedley and Kessler 1992; Simha and Ramaswamy 2002), this collectively gives rise to a directed contractile bulk stress, such that

$$\hat{\boldsymbol{\Sigma}}_{\rm out} - \hat{\boldsymbol{\Sigma}}_{\rm in} = \sigma \hat{\boldsymbol{I}} + \alpha \boldsymbol{n}\boldsymbol{n}, \tag{2.16}$$

with $\alpha > 0$ the magnitude of the directed contractile stress and $\boldsymbol{n} = (\cos\theta_{\rm SF}, \sin\theta_{\rm SF})$ the average direction of the stress fibers (Fig. 2.4c inset). The ratio between isotropic contractility σ and directed contractility α measures the degree of anisotropy of the bulk stresses. With this stress tensor the force balance equation (2.1) becomes:

$$\lambda' \boldsymbol{T} + (\sigma + \lambda\kappa)\boldsymbol{N} + \alpha(\boldsymbol{n}\cdot\boldsymbol{N})\boldsymbol{n} = \boldsymbol{0}. \tag{2.17}$$

Because $\boldsymbol{n}$ has, in general, non-vanishing projections on both the tangent and normal directions of the cell edge, this condition implies that in the presence of an anisotropic cytoskeleton, the cortical tension λ is no longer constant along the cell cortex.

It is useful to introduce a number of simplifications, with the goal of highlighting the physical mechanisms entailed in Eq. (2.17). As the orientation of the stress fibers varies only slightly along a single cellular arc, one can assume $\theta_{\rm SF}$ to be constant along each arc, but different, in general, from arc to arc. Moreover, as all the arcs share the same bulk, we consider the bulk stresses σ and α uniform throughout the cell. Let us then look at a specific cellular arc and, without loss of generality, choose to orient the cell in such a way the stress fibers are parallel to the $y-$axis. Thus $\vartheta_{\rm SF} = \pi/2$ (Fig. 2.5a). Using Eq. (2.2) and taking advantage of the fact that $\boldsymbol{n}$ does not change along the arc, one can express all terms

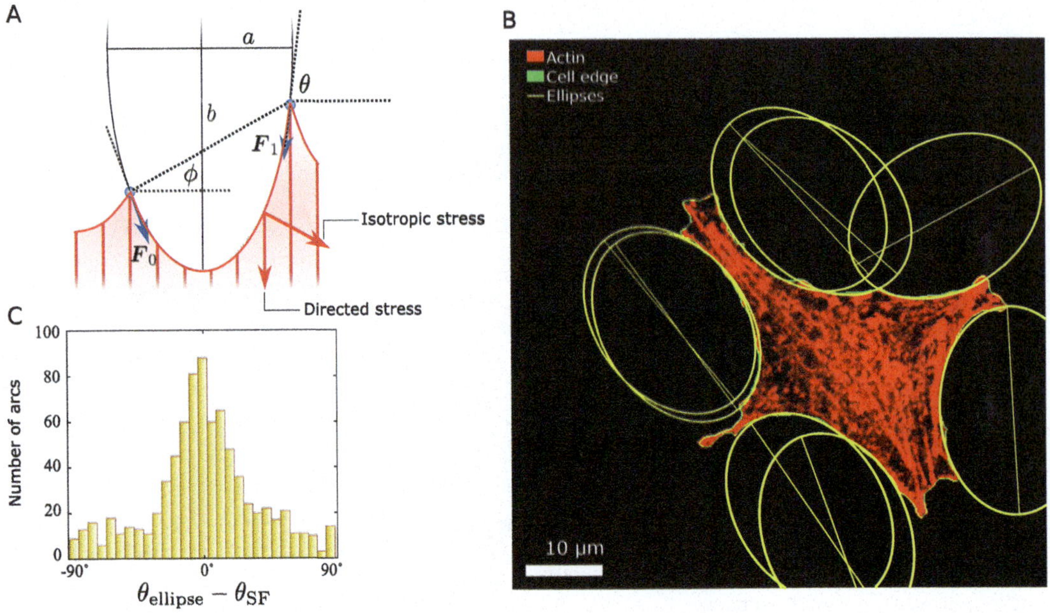

Fig. 2.5 The anisotropic cytoskeleton is reflected in the elliptical shape of the cell edge. (**a**) Schematic representation of our model for $\theta_{\rm SF} = \pi/2$. A force balance between isotropic stress, directed stress and line tension results in the description of each cell edge segment (red curve) as part of an ellipse of aspect ratio $a/b = \sqrt{\gamma}$, unique to each cell. The cell exerts forces $\boldsymbol{F}_0$ and $\boldsymbol{F}_1$ on the adhesion sites (blue). (**b**) An epithelioid cell (same cell as in Fig. 2.4) with a unique ellipse (yellow) fitted to its edges (green). The end-points of the arcs are identified based on the forces exerted on the pillars. The orientations of the major axes (yellow lines) are parallel to the local orientations of the stress fibers. Scalebar is 10 μm. (**c**) Histogram of $\theta_{\rm ellipse} - \theta_{\rm SF}$, with $\theta_{\rm ellipse}$ the orientation of the major axis of the fitted ellipse and $\theta_{\rm SF}$ the measured orientation of the stress fibers. The mean of this distribution is 0° and the standard deviation is 36°. (Reproduced from Pomp et al. 2018)

in Eq. (2.17) as total derivatives and integrate the equation directly. This yields:

$$\lambda \boldsymbol{T} + (\sigma \hat{\boldsymbol{I}} + \alpha \boldsymbol{nn}) \cdot \boldsymbol{r}^{\perp} = \boldsymbol{C}_1. \tag{2.18}$$

where $\boldsymbol{r}^{\perp} = (-y, x)$ and $\boldsymbol{C}_1 = (C_{1x}, C_{1y})$ is an integration constant. Then, using $\boldsymbol{n} = \hat{\boldsymbol{y}}$ and $\boldsymbol{T} = (\cos\theta, \sin\theta)$, we can simply this as:

$$\lambda \cos\theta = C_x + \sigma y \tag{2.19a}$$

$$\lambda \sin\theta = C_y - (\alpha + \sigma)x, \tag{2.19b}$$

from which, using $\tan\theta = dy/dx$ and integrating, we obtain a general solution of the force-balance equation in the form:

$$\frac{x^2}{\gamma} + y^2 - \frac{2C_{1y}}{\sigma} x + \frac{2C_{1x}}{\sigma} y = C_2, \tag{2.20}$$

where $\gamma = \sigma/(\alpha + \sigma)$ and C_2 is another integration constant. Notice that, if both σ and α are positive for a contractile system, $\gamma < 1$. Eq. (2.20) describes an ellipse whose minor and major semi-axes are $a = \sqrt{\gamma C_2}$ and $b = \sqrt{C_2}$ respectively and whose center is determined by $\boldsymbol{C}_1$. For simplicity, we can choose the origin of our reference frame to coincide with the center of the ellipse, so that $\boldsymbol{C}_1 = \boldsymbol{0}$. Using again Eq. (2.19) with $\tan\theta = -x/(\gamma y)$, we can further obtain an expression for the cortical tension as a function of the turning angle θ, namely:

$$\frac{\lambda^2}{\sigma^2} = C_2 \frac{1 + \tan^2\theta}{1 + \gamma \tan^2\theta}. \tag{2.21}$$

This expression highlights the physical meaning of the constant C_2. As the right-hand side attains its minimal values when $\theta = 0$, thus when tangent vector is perpendicular to the stress fibers, C_2 is related with the minimal tension $\lambda_{\min}$ withstood by the cortical actin, namely $C_2 = \lambda_{\min}^2/\sigma^2$, so that the shape of the cellular arc is described by the implicit equation:

$$\frac{\sigma^2}{\gamma \lambda_{\min}^2} x^2 + \frac{\sigma^2}{\lambda_{\min}^2} y^2 = 1, \tag{2.22}$$

and the tension withstood by the cortical actin is given, as a function of the turning angle, by:

$$\lambda = \lambda_{\min} \sqrt{\frac{1 + \tan^2\theta}{1 + \gamma \tan^2\theta}}. \tag{2.23}$$

In summary, in the presence of directed stresses the equilibrium conformation of the cell edge consists of arcs of an ellipse of semi-axes $a = \sqrt{\gamma}\,\lambda_{\min}/\sigma$ and $b = \lambda_{\min}/\sigma$ and whose major axis is parallel to the stress fibers. The dimensionless quantity γ highlights the interplay between the forces experienced by the cell edge and its shape: on the one hand, γ characterizes the anisotropy of the bulk stress, on the other hand it determines the anisotropy of the cell shape.

The key prediction the anisotropic tension model is illustrated in Fig. 2.5b, where the contour of the same cell shown in Fig. 2.4a has been fit with ellipses (Pomp et al. 2018). Whereas large variations in the circles' radii were required in Fig. 2.4a, a unique ellipse ($\gamma = 0.52$, $\lambda_{\min}/\sigma = 13.4\,\mu\text{m}$) faithfully describes all the arcs in the cell. Figure 2.5c shows the distribution of the difference between the orientation θ_{ellipse} of the major axis of the fitted ellipse and the measured orientation θ_{SF} of the stress fibers. The distribution peaks at 0° and has a width of 36°, demonstrating that the orientation of the ellipses is parallel, on average, to the local orientation of the stress fibers as predicted by the model.

The traction forces $\boldsymbol{F}_0$ and $\boldsymbol{F}_1$ can be straightforwardly calculated from Eqs. (2.17) and (2.23) in the form:

$$\frac{\boldsymbol{F}_0}{\lambda_{\min}} = (\beta \sin\phi + \zeta \cos\phi)\,\hat{\boldsymbol{x}} + \left(-\frac{\beta}{\gamma} \cos\phi + \zeta \sin\phi\right) \hat{\boldsymbol{y}}, \tag{2.24a}$$

$$\frac{\boldsymbol{F}_1}{\lambda_{\min}} = (\beta \sin\phi - \zeta \cos\phi)\,\hat{\boldsymbol{x}} + \left(-\frac{\beta}{\gamma} \cos\phi - \zeta \sin\phi\right) \hat{\boldsymbol{y}}, \tag{2.24b}$$

where:

$$\beta = \frac{d}{2b}, \qquad \zeta = \sqrt{\frac{1+\tan^2\phi}{1+\gamma\tan^2\phi} - \frac{\beta^2}{\gamma}}. \tag{2.25}$$

Here d is the distance between the positions of both forces on the ellipse, b is the major semi-axis of the ellipse and ϕ is the angle that the line through both points makes with the x−axis (see Fig. 2.5a).

2.4 The Effect of Bending Elasticity

The contour models reviewed in Sects. 2.2 and 2.3 postulate that the stresses arising in the actin cortex are purely contractile, thus tangential to the cell edge. Non-tangential stresses can build up as a consequence of the bending elasticity of the actin cortex as well as the plasma membrane. The effect of bending elasticity was considered in Banerjee and Giomi (2013) in order to account for the prominent polymorphism observed in experiments on cardiac myocytes adhering to substrates of varying stiffness (Chopra et al. 2011). In this work, myocytes grown on substrates having material properties mimicking physiological stiffness (5–10 kPa), were observed to spread less and develop convex and well rounded morphologies. In contrast, while plated on stiffer gels or glass, the same cell type is more likely to exhibit a concave shape and greater spread area. This crossover from convex to concave, in particular, cannot be explained from the simple tension model or the anisotropic tension model discussed in the previous sections, as these lack of passive restoring forces able to contrast the formation of the highly curved regions (i.e. kinks) that characterize a closed plane curve whose curvature κ is everywhere negative. In this respect, bending elasticity is the most natural choice among possible restoring mechanisms.

From a theoretical perspective, the problem arising by incorporating bending elasticity in a contour model of adherent cells, directly relates with another classic problem in mechanics: finding the shape of an infinitely long elastic pipe subject to uniform later pressure. This problem was formulated by Maurice Lévy in 1884 (Lévy 1884) and for over a century drew the attention of many researchers, due to its tremendous richness of polymorphic and multistable solutions (Arreaga et al. 2002; Djondjorov et al. 2011; Flaherty et al. 1972; Giomi 2013; Giomi and Mahadevan 2012; Mora et al. 2012; Tadjbakhsh and Odeh 1967; Vassilev et al. 2008). Unlike the classic Lévy problem, however, the model proposed here for adhering cells does not involve any constraint on the length of the boundary, which is then only solely constrained by the adhesion with the substrate (Giomi 2013). This feature, introduces in the model a number of crucial mechanical properties, including an adaptive bending stiffness of the cell boundary.

As it is known from classical elasticity of rods (see e.g. Landau and Lifshitz 1970), a slender structure forced to bend on the plane, is subject to a moment resultant

$$\boldsymbol{M} = B\kappa\hat{\boldsymbol{z}}. \tag{2.26}$$

with B the bending stiffness and κ the curvature, with the usual sign convention introduced in Sect. 2.1. Furthermore, the local balance of bending moments (Landau and Lifshitz 1970) requires:

$$\boldsymbol{M}' + \boldsymbol{T} \times \boldsymbol{F} = \boldsymbol{0}. \tag{2.27}$$

Now, taking $\boldsymbol{F} = F_T\boldsymbol{T} + F_N\boldsymbol{N}$, with F_T and F_N the tangential and normal components of the stress resultant, and using $\boldsymbol{T} \times \boldsymbol{N} = \hat{\boldsymbol{z}}$, one can cast Eqs. (2.1) and (2.27) into:

$$B\kappa' + F_N = 0, \tag{2.28a}$$

$$F_T' - \kappa F_N + \boldsymbol{T} \cdot (\hat{\boldsymbol{\Sigma}}_{\text{out}} - \hat{\boldsymbol{\Sigma}}_{\text{in}}) \cdot \boldsymbol{N} + \boldsymbol{T} \cdot \boldsymbol{f}_{\text{ext}} = 0, \tag{2.28b}$$

$$\kappa F_T + F_N' + \boldsymbol{N} \cdot (\hat{\boldsymbol{\Sigma}}_{\text{out}} - \hat{\boldsymbol{\Sigma}}_{\text{in}}) \cdot \boldsymbol{N} + \boldsymbol{N} \cdot \boldsymbol{f}_{\text{ext}} = 0. \tag{2.28c}$$

To make progress we restrict ourselves to cells with discrete rotational symmetry (Fig. 2.6a) and isotropic cytoskeleton. In this case one can assume $\boldsymbol{f}_{\text{ext}} = f_{\text{ext}}\boldsymbol{N}$. Then, Eqs. (2.7), (2.28a) and

(2.28b) yield:

$$F = \left(\lambda - \frac{1}{2}B\kappa^2\right)T - B\kappa' N, \quad (2.29)$$

whereas Eq. (2.28c) yields an equation for the curvature κ:

$$B\left(\kappa'' + \frac{1}{2}\kappa^3\right) - \lambda\kappa - \sigma - f_{\text{ext}} = 0. \quad (2.30)$$

For $f_{\text{ext}} = \text{const}$ this is the equation dictating the shape of an infinitely long pipe subject to a uniform pressure, or, alternatively, of a planar *elastica* spanned by a capillary film (Giomi and Mahadevan 2012; Mora et al. 2012). Unlike these examples, however, the length of the cell edge is not fixed and shall be determined from the balance between contractility, bending elasticity and the elastic response of the substrate embodied in the force per unit length f_{ext}.

To illustrate this last point, we can consider the simplified case in which peripheral contractility is negligible (i.e. $\lambda \approx 0$) and the cell periphery continuously adhere to the substrate along the cell edge. As there is no special direction on the plane, we can assume the cell to be a circle of radius R centered at the origin. The force per unit length resulting form the deformation of the substrate, can then be expressed in the simple form:

$$f_{\text{ext}} = -\frac{k_s(R - R_0)}{\mathcal{L}}, \quad (2.31)$$

where k_s is the elastic stiffness of the substrate, R_0 is the radius of the cell before this starts stretching the substrate (i.e. once adhesions are formed and contractile forces start to build up) and $\mathcal{L} = 2\pi R$ is the cell perimeter. Thus, setting $\kappa = 1/R$ in Eq. (2.30) yields the following cubic equation:

$$(k_s + 2\pi\sigma)R^3 - k_s R_0 R^2 - \pi B = 0, \quad (2.32)$$

The equation contains two length scales, R_0 and $\xi = (B/\sigma)^{1/3}$, and a dimensionless control parameter k_s/σ expressing the relative amount of adhesion and contraction. For very soft anchoring $k_s \ll \sigma$ and Eq. (2.32) admits the solution $R = \xi/2^{1/3}$. Thus non-adherent cells or cells adhering to extremely soft substrates (i.e. $k_s = 0$), are predicted to have a radius of curvature that scales as $R \sim \sigma^{-1/3}$. The same scaling law is also predicted using *active cable network* models of an adherent cell (Torres et al. 2012). If the cell is rigidly pinned at adhesion sites, $k_s \gg \sigma$ and $R \to R_0$. For intermediate values of k_s/σ the optimal radius R interpolates between ξ and R_0 and is an increasing function of the substrate stiffness k_s, in case $\xi < R_0$, or a decreasing function if $\xi > R_0$. For $\xi = R_0$, the lower and upper bound coincide, and the solution is $R = R_0$. In particular, the case $R_0 > \xi$ reproduces the experimentally observed trend that cell projected area increases with increasing substrate stiffness before reaching a plateau at higher stiffnesses (Chopra et al. 2011; Engler et al. 2004; Yeung et al. 2005). The asymptotic behavior and various limits of the solution are well captured by the interpolation formula:

$$R \approx \frac{k_s R_0 + 6\pi\sigma\,\xi}{k_s + 6\pi\sigma} \quad (2.33)$$

indicating that larger surface tension, hence larger cell contractility σ leads to lesser spread area, consistent with the experimental observation that myosin-II activity retards the spreading of cells (Wakatsuki et al. 2003). Standard stability analysis of this solution under a small periodic perturbation in the cell radius shows that the circular shape is always stable for any values of the parameters σ, k_s and R_0.

For cells adhering to discrete number of adhesion sites, one can show that the circular solution for the cell boundary is never stable. For simplicity, we assume that N_A adhesion sites are located at the vertices of a regular polygon of circumradius R_0 (Fig. 2.6a). The force per unit length exerted by the substrate is then:

$$f_{\text{ext}} = -k_s \sum_{i=0}^{N_A - 1} \delta(s - iL)[r(s) - R_0], \quad (2.34)$$

where L the distance between consecutive adhesion points and $r(s) = |\boldsymbol{r}(s)|$. Using again the

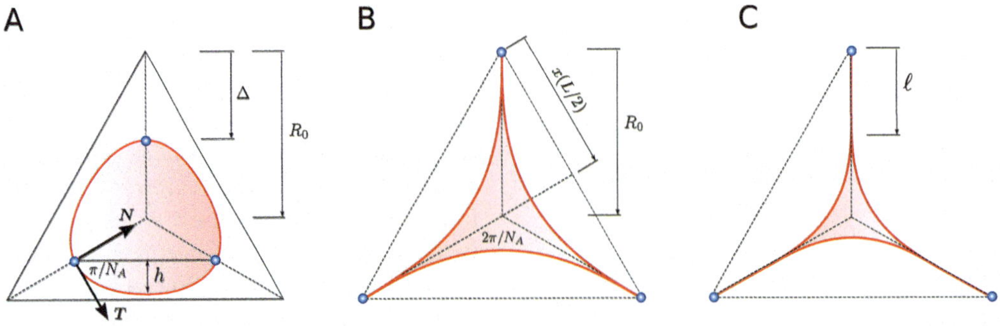

Fig. 2.6 Cell anchored onto three pointwise adhesions located at the vertices of an equilateral triangle. (**a**) For small contractility values, the cell contour is everywhere convex with constant width. (**b**) When the contractility reaches a critical value σ_0, cell contour is purely concave with cusps at adhesion points. (**c**) For $\sigma > \sigma_0$, cusps gives rise to protrusion of length ℓ. (Adapted from Banerjee and Giomi 2013)

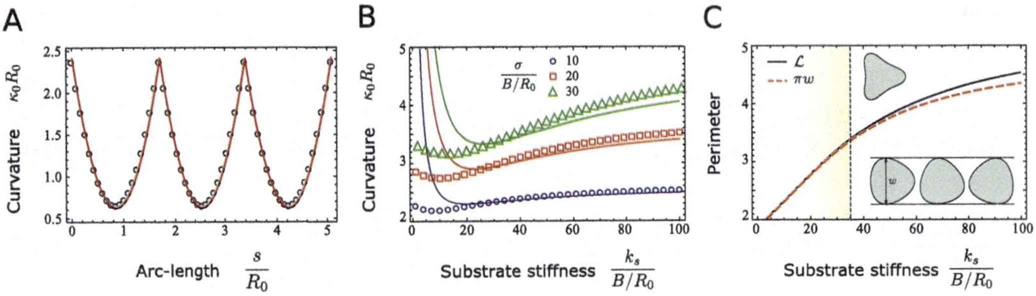

Fig. 2.7 Cell anchored onto three pointwise adhesions located at the vertices of an equilateral triangle. (**a**) Curvature versus arc-length for $\sigma R_0^3/B = 10$, $k_s R_0^3 = 50$ and $N_A = 3$. The circles are obtained from a numerical solution of Eq. (2.30), while the solid lines corresponds to our analytical approximation. (**b**) The end-point curvature κ_0 at the adhesion points as a function of the substrate stiffness for various contractility values. The points are obtained from numerical simulations while the solid lines correspond to our analytical approximation. (**c**) The total cell length $\mathcal{L}$ as a function of adhesion stiffness. For small stiffnesses the cell boundary form a curve of constant width (lower inset) and $\mathcal{L} = \pi w$, with w the width of the curve. This property breaks down for larger stiffnesses when inflection points develops (upper inset). (Reproduced from Banerjee and Giomi 2013)

rotational symmetry of the problem, we can assume that the substrate is stretched at all adhesion points by the same amount. Then $r(s) - R_0 = \Delta = \text{const}$ and Eq. (2.30) reduces to:

$$B\left(\kappa'' + \frac{1}{2}\kappa^3\right) - \lambda\kappa - \sigma + k_s\Delta \sum_{i=0}^{N_A-1} \delta(s - iL) = 0. \tag{2.35}$$

Integrating Eq. (2.35) along an infinitesimal neighborhood of the i-th adhesion point, one finds the following condition for the derivative of the curvature at the adhesion points:

$$\kappa_i' = -\frac{k_s}{2B}\,\Delta. \tag{2.36}$$

The local curvature of the segment lying between adhesion points is, on the other hand, determined by the equation:

$$\kappa'' + \frac{1}{2}\kappa^3 - \frac{\lambda}{B}\kappa - \frac{\sigma}{B} = 0, \tag{2.37}$$

with the boundary conditions : $\kappa(iL) = \kappa((i+1)L) = \kappa_0$, with $i = 1, 2 \ldots N_A$ and κ_0 a constant to be determined. Without loss of generality we consider a segment located in the interval $0 \le s \le L$. Although an exact analytic solution of this nonlinear equation is available (e.g. Banerjee and Giomi 2013), an excellent approximation can be obtained by neglecting the cubic nonlinearity (Fig. 2.7a). With this simplification, Eq. (2.35) admits a simple quadratic solution of the form:

$$\kappa(s) = \kappa_0 + \frac{\sigma}{2B} s(s-L). \tag{2.38}$$

Eqs. (2.38) and (2.36) immediately allow us to derive a condition on the cell perimeter, namely

$$L = \frac{k_s \Delta}{\sigma}. \tag{2.39}$$

This leads, furthermore, to a linear relation between traction force $F_{\text{traction}} = k_s \Delta$, and cell size:

$$F_{\text{traction}} = \sigma L, \tag{2.40}$$

which is indeed observed in traction force measurements on large epithelial cells (Mertz et al. 2012).

To determine the end-point curvature κ_0, one can use the turning tangents theorem for a simple closed curve, Eq. (2.4), which requires $\int_0^L \mathrm{d}s\, \kappa = 2\pi/N_A$. This leads to following relation between local curvature and segment length, or equivalently traction force, at the adhesion sites:

$$\kappa_0 = \frac{\sigma L^2}{12B} + \frac{2\pi}{N_A L}. \tag{2.41}$$

A plot of κ_0 as a function of the substrate stiffness is shown in Fig. 2.7b. Finally, to determine the optimal length of the cell segment L, we are going to make use of a remarkable geometrical property of the curve obtained from the solution of Eq. (2.35) with discrete adhesions: the fact of being a *curve of constant width* (Gray 1997). The width of a curve is the distance between the uppermost and lowermost points on the curve (see lower inset of Fig. 2.7c). In general, such a distance depends on how the curve is oriented. There is however a special class of curves, where the width is the same regardless of their orientation. The simplest example of a curve of constant width is clearly a circle, in which case the width coincides with the diameter. A fundamental property of curves of constant width is given by the Barbier's theorem (Gray 1997), which states that the perimeter $\mathcal{L}$ of any curve of constant width is equal to width w multiplied by π: $\mathcal{L} = \pi w$. As illustrated in Fig. 2.7c, this is confirmed by numerical simulations for low to intermediate values for contractility and stiffness. With our setting, the cell width is given by:

$$w = (R_0 - \Delta)\left(1 + \cos\frac{\pi}{N_A}\right) + h\left(\frac{L}{2}\right), \tag{2.42}$$

where $h(s) = \int_0^s \mathrm{d}s' \sin\theta(s')$ is the height of the curve above a straight line between two adhesions points (Fig. 2.7a) and

$$\theta(s) = \int_0^s \mathrm{d}s'\, \kappa(s') = \theta_0 + \kappa_0 s + \frac{\sigma}{12B} s^2(2s - 3L) \tag{2.43}$$

the usual turning angle (Fig. 2.1b). For small angles h can be approximated as :

$$h(s) \approx s(L-s)\left[\frac{\pi}{N_A L} - \frac{\sigma}{12B} s(L-s)\right] \tag{2.44}$$

Using this, Eq. (2.42) and the Barbier's theorem with $\mathcal{L} = N_A L$ allow us to obtain a quartic equation for the cell length:

$$\frac{N_A L}{\pi} = \left(1 + \cos\frac{\pi}{N_A}\right)\left(R_0 - \frac{\sigma L}{2k_s}\right) + \frac{L}{4}\left(\frac{\pi}{N_A} - \frac{1}{96}\frac{\sigma L^3}{B}\right). \tag{2.45}$$

Figure 2.8 shows plots of the traction $F_{\text{traction}} = \sigma L$ with L determined by solving Eq. (2.45). These results support the experimental trend that traction force increases monotonically with substrate stiffness k_s before plateauing to a finite value for higher stiffnesses (Ghibaudo et al. 2008; Mitrossilis et al. 2009). The plateau value increases with increasing contractility (Fig. 2.8a). Traction force grows linearly with increasing contractility for $\sigma R_0^3/B \ll 1$, before saturating to the value $k_s R_0$ at large contractility $\sigma R_0^3/B \gg 1$, as shown in Fig. 2.8b. Equation (2.13) is also consistent with experimentally observed trend that reducing contractility by increasing the dosage of myosin inhibitor Blebbistatin, leads to monotonic drop in traction forces (Mitrossilis et al. 2009).

For low to intermediate values of σ and k_s, cell shape is convex and has constant width. Upon

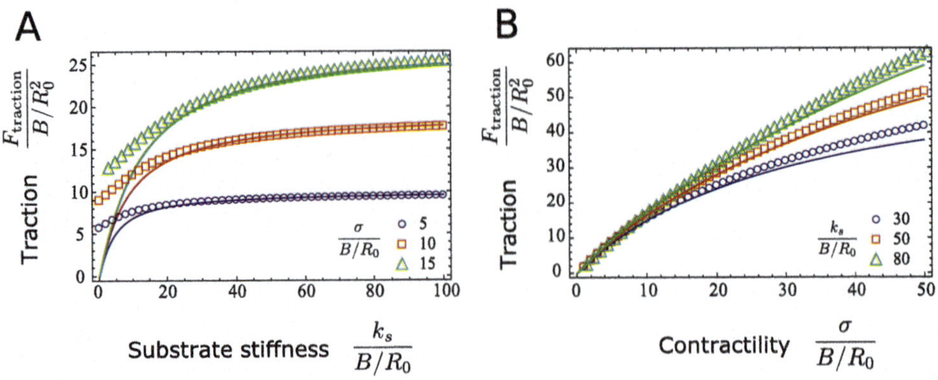

Fig. 2.8 Traction force as a function of substrate stiffness (**a**) and contractility (**b**) obtained from a numerical solution of Eq. (2.35). Solid curves denote the approximate traction values obtained from Eq. (2.45). (Adapted from Banerjee and Giomi 2013)

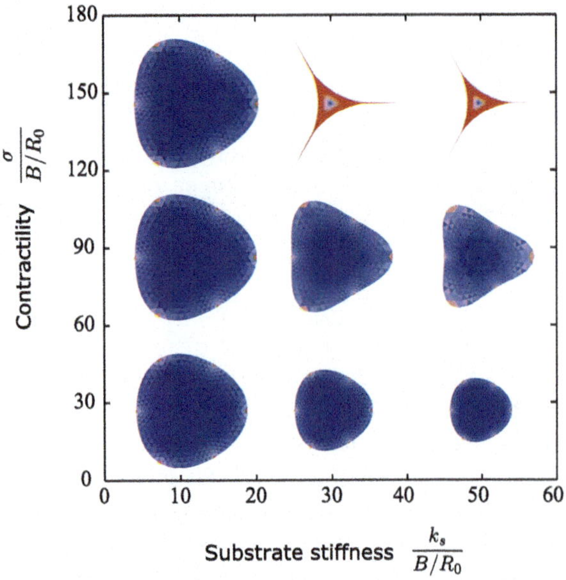

Fig. 2.9 Phase diagram in σ-k_s plane showing optimal configuration obtained by numerical minimization of the energy (2.10) for $N_A = 3$. (Adapted from Banerjee and Giomi 2013)

increasing σ above a k_s−dependent threshold, however, the cell boundary becomes inflected (see Fig. 2.9 and upper inset of Fig. 2.7c). Initially, a region of negative curvature develops in proximity of the mid point between two adhesions, but as the surface tension is further increased, the size of this region grows until positive curvature is preserved only in a small neighborhood of the adhesion points. Due to the presence of local concavities, the cell boundary is no longer a curve of constant width.

If σ is further increased, the inflected shape collapses giving rise to the star-shaped configurations shown in upper right corner of Fig. 2.9. These purely concave configurations are made by arcs whose ends meet in a cusp. The cusp is then connected to the substrate by a protrusion consisting of a straight segment stretching until the adhesion point rest position, so that $\Delta = 0$ (Fig. 2.6c) and, consistent with Eq. (2.36), $\kappa = \text{const} = 0$ at the adhesion point. Unlike the previous transition from convex to non-convex shapes, this second transition occurs discontinuously and is accompanied by a region of bistability (see Banerjee and Giomi (2013) for further detail). Away from the protrusion, the curvature has still the form given in Eq. (2.38), with $\kappa_0 = 0$ so that the boundary is everywhere concave or flat and the bending moment $\boldsymbol{M} = B\kappa\hat{\boldsymbol{z}}$ does not experience any unphysical discontinuity at the protrusions origin.

The length of the protrusion can be readily obtained using a similarity transformation. This construction, first investigated by Flaherty et al. (1972) for the original Lévy problem, relies on the invariance of Eq. (2.37) under the following scaling transformation:

$$\{s, \kappa, \lambda, \sigma\} \to \left\{\Lambda s, \frac{\kappa}{\Lambda}, \frac{\lambda}{\Lambda^2}, \frac{\sigma}{\Lambda^3}\right\}. \tag{2.46}$$

with Λ a scaling factor. Calling then σ_0 the value of σ at which the protrusion have zero length, the shape of the cell edge at any $\sigma > \sigma_0$ can be constructed starting from the reference shape illustrated in Fig. 2.6b as follows. First one calculates the scaling factor $\Lambda = (\sigma_0/\sigma)^{1/3}$ associated with the new σ value. The reference shape is then rescaled by Λ in such a way that the cusps are now disconnected from the original adhesion points. Finally, the cusps and the adhesion points are reconnected by straight segments of length $\ell_{\rm p} = R_0(1 - \Lambda)$ (since R_0 is the circumradius of the reference shape and ΛR_0 that of the rescaled shape). This latter step, ultimately allows us to formulate a scaling law for the length of protrusions, namely:

$$\frac{\ell}{R_0} = 1 - \left(\frac{\sigma_0}{\sigma}\right)^{\frac{1}{3}}. \tag{2.47}$$

This transition from a smooth shape to a self-contacting shape with cusps is reminiscent of the post-buckling scenario of an elastic ring subject to a uniform pressure, but unlike this case, where the system undergoes a continuous transition from a simple curve to a curve with lines of contact, here the transition is discontinuous along both the loading branch (increasing k_s) and the unloading branch (decreasing k_s). The transition has moreover a strong topological character since it involves a jump in the rotational index of the curve, whose total curvature after the transition becomes:

$$\oint_{\partial M} ds\, \kappa = \pi(2 - N_A),$$

by virtue of Eq. (2.5). Some further detail about the geometry of protrusions in this model can be found in Banerjee and Giomi (2013).

2.5 Concluding Remarks

In this chapter we have reviewed a simple theoretical framework for modeling the mechanical aspects of cell-substrate interaction. Several experimental works have demonstrated the latter to play a critical role in regulating a variety of cellular processes, from morphogenesis, motility, to cell linage and fate. While adhering to the extracellular matrix, cell develop specific morphologies depending on the geometrical and mechanical properties of their micro-environment. In turn, the forces arising inside the cell in response to this structural reorganization, drive biochemical cascades that not only feedback on the mechanical cell-matrix interaction, but also influence other processes such as cell cycle control and differentiation. Whereas the extraordinary complexity of these biomechanical pathways is still elusive, much progress has been made in understanding how the presence of actively generated contractile forces, on the one hand, and the absence of hard geometric constraints, on the other, give rise to a whole new class of mechanical phenomena commonly found in living systems, such as adaptivity, polymorphism, multistability etc. The simple contour models reviewed in this chapter, have given an important contribution in this respect, as they allow to cast the problem in a form that is often analytically tractable, thanks to the reduced dimensionality.

Whereas insightful under many respects, contour models do not give access to the dynamics of the molecular process involved in cell adhesion and migration, such as the association dissociation events in the adhesion clusters or the regulation of myosin expression in the presence of various mechanical cues. This create the demand for a more comprehensive theoretical framework, where the continuum mechanics standpoint of contour models could be integrated into a multi-scale approach able to account for the mechanical, biochemical and genetic aspects of cellular organization alike.

Acknowledgements I am indebted with Koen Schakenraad, Thomas Schmidt, Wim Pomp and Shiladitya Banerjee for contributing to the work reviewed here. This work is partially supported by the Netherlands Organisation for Scientific Research (NWO/OCW), as part of the Frontiers of Nanoscience program and the Vidi scheme.

References

Arreaga G, Capovilla R, Chryssomalakos C, Guven J (2002) Area-constrained planar elastica. Phys Rev E 65:031801
Banerjee S, Giomi L (2013) Polymorphism and bistability in adherent cells. Soft Matter 9:5251–5260
Barry NP, Bretscher MS (2010) Dictyostelium amoebae and neutrophils can swim. Proc Natl Acad Sci U S A 107:11376–11380
Bar-Ziv R, Tlusty T, Moses E, Safran SA, Bershadsky AD (1999) Pearling in cells: a clue to understanding cell shape. Proc Natl Acad Sci USA 96:10140–10145
Bischofs IB, Klein F, Lehnert D, Bastmeyer M, Schwarz US (2008) Filamentous network mechanics and active contractility determine cell and tissue shape. Biophys J 95:3488–3496
Bischofs IB, Schmidt SS, Schwarz US (2009) Effect of adhesion geometry and rigidity on cellular force distributions. Phys Rev Lett 103:048101
Burridge K, Chrzanowska-Wodnicka M (1996) Focal adhesions, contractility and signalling. Annu Rev Cell Dev Biol 12:463–519
Burridge K, Wittchen ES (2013) The tension mounts: stress fibers as force-generating mechanotransducers. J Cell Biol 200:9–19
Chopra A, Tabdanov E, Patel H, Janmey PA, Kresh JY (2011) Cardiac myocyte remodeling mediated by N-cadherin-dependent mechanosensing. Am J Physiol Heart Circ Physiol 300:H1252-H1266
Danen EHJ, Sonneveld P, Brakebusch C, Fässler R, Sonnenberg A (2002) The fibronectin-binding integrins $\alpha 5\beta 1$ and $\alpha v\beta 3$ differentially modulate RhoA-GTP loading, organization of cell matrix adhesions, and fibronectin fibrillogenesis. J Cell Biol 159:1071–1086
DeTurck D, Gluck H, Pomerleano D, Vick DS (2007) The four vertex theorem and its converse. Not Am Math Soc 54:192–207
Discher DE, Janmey PA, Wang YL (2005) Tissue cells feel and respond to the stiffness of their substrate. Science 310:1139–1143
Djondjorov P, Vassilev V, Mladenov I (2011) Analytic description and explicit parametrisation of the equilibrium shapes of elastic rings and tubes under uniform hydrostatic pressure. Int J Mech Sci 53:355–364
do Carmo MP (1976) Differential geometry of curves and surfaces. Prentice-Hall, Upper Saddle River
Engler A, Bacakova L, Newman C, Hategan A, Griffin M, Discher D (2004) Substrate compliance versus ligand density in cell on gel responses. Biophys J 86:617–628
Engler AJ, Sen S, Sweeney HL, Discher DE (2006) Matrix elasticity directs stem cell lineage specification. Cell 126:677–689
Flaherty J, Keller J, Rubinow S (1972) Post buckling behavior of elastic tubes and rings with opposite sides in contact. SIAM J Appl Math 23:446–455
Geiger B, Spatz JP, Bershadsky AD (2009) Environmental sensing through focal adhesions. Nat Rev Mol Cell Biol 10:21–33
Ghibaudo M, Saez A, Trichet L, Xayaphoummine A, Browaeys J, Silberzan P, Buguin A, Ladoux B (2008) Traction forces and rigidity sensing regulate cell functions. Soft Matter 4:1836–1843
Ghibaudo M, Trichet L, Le Digabel J, Richert A, Hersen P, Ladoux B (2009) Substrate topography induces a crossover from 2D to 3D behavior in fibroblast migration. Biophys J 97:357–368
Giomi L (2013) Softly constrained films. Soft Matter 9:8121–8139
Giomi L, Mahadevan L (2012) Minimal surfaces bounded by elastic lines. Proc R Soc A 468:1851–1864
Gray A (1997) Modern differential geometry of curves and surfaces with mathematica. CRC-Press, Boca Raton
Janmey PA, McCulloch CA (2007) Cell mechanics: integrating cell responses to mechanical stimuli. Annu Rev Biomed Eng 9:134
Jülicher F, Kruse K, Prost J, Joanny JF (2007) Active behavior of the cytoskeleton. Phys Rep 449:3–28
Landau LD, Lifshitz EM (1970) Theory of elasticity. A course of theoretical physics, vol 7. Pergamon Press, Oxford
Lévy M (1884) Mémoire sur un nouveau cas intégrable du problème de lélastique et l'une des ses applications. J Math Pure Appl 10:5–42
Lo CM, Wang HB, Dembo M, Wang YL (2000) Cell movement is guided by the rigidity of the substrate. Biophys J 79:144–152
Mendez MG, Janmey PA (2012) Transcription factor regulation by mechanical stress. Int J Biochem Cell Biol 44:728–732
Mertz AF, Banerjee S, Che Y, German GK, Xu Y, Hyland C, Marchetti MC, Horsley V, Dufresne ER (2012) Scaling of traction forces with the size of cohesive cell colonies. Phys Rev Lett 108:198101
Midwood KS, Williams LV, Schwarzbauer JE (2004) Tissue repair and the dynamics of the extracellular matrix. Int J Biochem Cell Biol 36:1031–1037
Mitrossilis D, Fouchard J, Guiroy A, Desprat N, Rodriguez N, Fabry B, Asnacios A (2009) Single-cell response to stiffness exhibits muscle-like behavior. Proc Natl Acad Sci USA 106:18243–18248

Mora S, Phou T, Fromental JM, Audoly B, Pomeau Y (2012) Shape of an elastic loop strongly bent by surface tension: experiments and comparison with theory. Phys Rev E 86:026119

Pedley TJ, Kessler JO (1992) Hydrodynamic phenomena in suspensions of swimming microorganisms. Annu Rev Fluid Mech 24:313–358

Pellegrin S, Mellor H (2007) Actin stress fibres. J Cell Sci 120:3491–3499

Pomp W, Schakenraad K, Balcıoğlu HE, van Hoorn H, Danen EHJ, Merks RMH, Schmidt T, Giomi L (2018) Cytoskeletal anisotropy controls geometry and forces of adherent cells. Phys Rev Lett 121:178101

Reinhart-King CA, Dembo M, Hammer DA (2008) Cell-cell mechanical communication through compliant substrates. Biophys J 95:6044–6051

Sabass B, Gardel ML, Waterman CM, Schwarz US (2008) High resolution traction force microscopy based on experimental and computational advances. Biophys J 94:207–220

Sawada Y, Tamada M, Dubin-Thaler BJ, Cherniavskaya O, Sakai R, Tanaka S, Sheetz MP (2006) Force sensing by mechanical extension of the Src family kinase substrate p130Cas. Cell 127:1015–1026

Schwarz US, Safran SA (2013) Physics of adherent cells. Rev Mod Phys 85:1327

Simha RA, Ramaswamy S (2002) Hydrodynamic fluctuations and instabilities in ordered suspensions of self-propelled particles Phys Rev Lett 89:058101

Sochol RD, Higa AT, Janairo RRR, Li S, Lin L (2011) Unidirectional mechanical cellular stimuli via micropost array gradients. Soft Matter 7:4606–4609

Tadjbakhsh I, Odeh F (1967) Equilibrium states of elastic rings. J Appl Math Anal Appl 18:59–74

Tan JL, Tien J, Pirone DM, Gray DS, Bhadriraju K, Chen CS (2003) Cells lying on a bed of microneedles: an approach to isolate mechanical force. Proc Natl Acad Sci USA 100:1484–1489

Tanimoto H, Sano M (2012) Dynamics of traction stress field during cell division. Phys Rev Lett 109:248110

Torres PG, Bischofs I, Schwarz U (2012) Contractile network models for adherent cells. Phys Rev E 85:011913

Trappmann B, Gautrot JE, Connelly JT, Strange DGT, Li Y, Oyen ML, Cohen Stuart MA, Boehm H, Li B, Vogel V, Spatz JP, Watt FM, Huck WTS (2012) Extracellular-matrix tethering regulates stem-cell fate. Nat Mater 11:642–649

Vassilev V, Djondjorov P, Mladenov I (2008) Cylindrical equilibrium shapes of fluid membranes. J Phys A: Math Gen 41:435201

Wakatsuki T, Wysolmerski R, Elson E (2003) Mechanics of cell spreading: role of myosin II. J Cell Sci 116:1617–1625

Yeung T, Georges PC, Flanagan LA, Marg B, Ortiz M, Funaki M, Zahir N, Ming W, Weaver V, Janmey PA (2005) Effects of substrate stiffness on cell morphology, cytoskeletal structure, and adhesion. Cell Motil Cytoskeleton 60:24–34

Force and Collective Epithelial Activities

3

Aldo Ferrari and Costanza Giampietro

Abstract

Cells apply forces to their surroundings to perform basic biological activities, including division, adhesion, and migration. Similarly, cell populations in epithelial tissues coordinate forces in physiological processes of morphogenesis and repair. These activities are highly regulated to yield the correct development and function of the body. The modification of this order is at the onset of pathological events and malfunctions. Mechanical forces and their translation into biological signals are the focus of an emerging field of research, shaping as a central discipline in the study of life and gathering knowledge at the interface of engineering, physics, biology and medicine. Novel engineering methods are needed to complement the classic instruments developed by molecular biology, physics and medicine. These should enable the measurement of forces at the cellular and multicellular level, and at a temporal and spatial resolution which is fully compatible with the ranges experienced by cells *in vivo*.

Keywords

Traction force microscopy · Collective migration · Epithelia · Acto-myosin contractility · Nano-printing

3.1 From Mesenchymal Migration to Collective Motion

Differentiated epithelial layers in the human body generate a continuous barrier that lines the inner and outer surface of all organs (e.g. lungs, mammary glands, intestine, and kidneys) and body cavities, including the luminal side of blood and lymphatic vessels (i.e. the endothelium). They exert a filtering activity that ensures protection, separation, sensing and transport. In quiescent epithelia, each constituent cell performs its function while anchored to the local basal matrix and to the neighboring cells, and therefore with relatively little movements. However, epithelial cells retain the ability to migrate to great distances, which can be unleashed by a breach in the monolayer continuity as in wound healing, or by the emergence of malignant behavior as in car-

A. Ferrari (✉)
EMPA, Swiss Federal Laboratories for Material Science and Technologies, Dübendorf, Switzerland

Laboratory of Thermodynamics in Emerging Technologies, ETHZ, Zurich, Switzerland

Institute for Mechanical Systems, ETH Zurich, Zürich, Switzerland
e-mail: aferrari@ethz.ch

C. Giampietro
Laboratory of Thermodynamics in Emerging Technologies, ETHZ, Zurich, Switzerland

C. A. M. La Porta, S. Zapperi (eds.), *Cell Migrations: Causes and Functions*, Advances in Experimental Medicine and Biology 1146,
https://doi.org/10.1007/978-3-030-17593-1_3

cinoma (Friedl and Gilmour 2009). Importantly, when epithelial cells part of a confluent layer migrate, they do so as a collective which moves in a coordinated and coherent fashion (Holmes 1914; Arboleda-Estudillo et al. 2010). The emergence of collective migration is orchestrated by a number of physical and molecular factors with complex interactions. Such intrinsic complexity is the reason why an integrative paradigm for collective migration remains elusive (Park et al. 2016).

The migration of isolated epithelial cells across flat substrates has been the subject of successful investigations, which led to the definition of a general pattern (Ridley et al. 2003; Lauffenburger and Horwitz 1996). Mesenchymal migration (Fig. 3.1) can be conceptualized as a sequential process involving four steps:

1. Initial cell polarization driven by localized actin polymerization which is then followed by the extension of a leading edge.
2. Once the leading edge contacts the extracellular matrix (ECM) components coating the surface, specific adhesion receptors of the integrin family start to cluster inducing the formation, stabilization and maturation of cell-substrate adhesions.
3. At this point actomyosin-mediated cell contractility tests the relative strength of adhesions established along the cell.
4. The loss of weak posterior adhesions is followed by retraction of the rear edge and net forward movement of the cell.

The establishment of this *in vitro* model of migration paved the way to fundamental studies revealing the biological mechanisms involved (Hynes 1992). In the last decade, with the help of the advancement of patterning nano and micro-technologies, it was also possible to demonstrate that, besides the well characterized biological signaling, a spatially-overlapping but distinct set of guidance cues is encoded by the physical properties of the substrate (Ferrari and Cecchini 2011). The mechanical rigidity, the density of adhesion points and the surface topography are independently read by cells in a process requiring a direct interaction with the surrounding microenvironment and thus termed contact guidance (Geiger et al. 2009).

This knowledge served as a platform to identify the role of cellular machineries establishing contact with the substrate (the integrin adhesions; (Geiger et al. 2009)), generating contractile forces (the actomyosin complex; (Ji et al. 2008; Petrie and Yamada 2015)) or responding to molecular (gradients of soluble molecules; (Keller 2012)) or physical (i.e. density of adhesion points (Arnold et al. 2008), rigidity (Lo et al. 2000), surface topography (Biela et al. 2009)) directional stimuli. In addition, the established model of mesenchymal migration provided a reference to resolve alternative migration modalities, which are adopted by specific cell types (Lammermann and Sixt 2009; Paluch et al. 2016).

In particular, a number of studies in both normal and tumor cells (Lo et al. 2000; Pel-

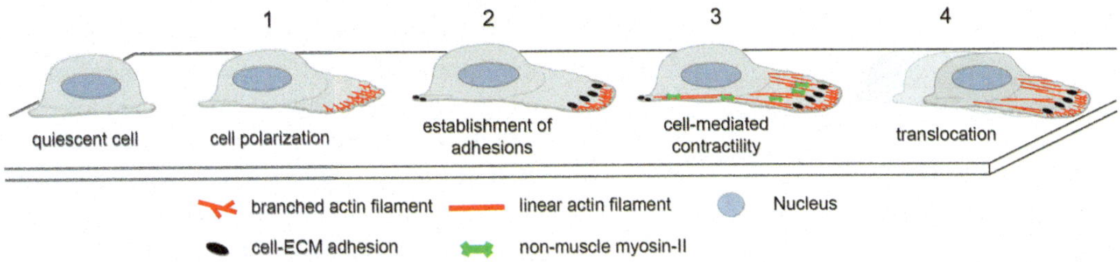

Fig. 3.1 Mesenchymal migration across 2D substrates. (*1*). Cell polarization is obtained through the extension of an actin-supported membrane protrusion at the leading edge. (*2*). The protrusion contacts the substrate and strong adhesions develop at the leading edge, while weak adhesions are established at the rear. (*3*). Non-muscle myosin-II accumulates along actin filaments and generates contractility along the cell body. (*4*). Adhesions at the rear are lost, resulting in a net forward movement of the cell

ham Jr and Wang 1997; Tzvetkova-Chevolleau et al. 2008; Wozniak et al. 2004; Zeng et al. 2006) revealed that increasing the substrate stiffness within a specific range (i.e. elastic modulus variations between 10 and 10,000 kPa; (Ferrari and Cecchini 2011)) induces a phenotype characterized by the stabilization of cell-substrate adhesions and by the activation of actomyosin contractility. Additionally, gradients in the elastic modulus drive migration toward regions of higher stiffness (durotaxis), implying that specific cell machineries act as mechanosensors during migration (Geiger et al. 2009; Lo et al. 2000). Similarly, the development of techniques to precisely control the number and density of adhesion points (Glass et al. 2003) provided much information regarding the establishment and maturation of integrin-based adhesions and revealed that cells polarize and migrate toward areas of higher ligand density (distance between adhesion points ranging from 15 to 250 nm; (Arnold et al. 2008, 2009; Cavalcanti-Adam et al. 2007)). Finally, the application of lithographic techniques, and particularly of soft and imprint lithography, yielded the generation of biocompatible substrates bearing textures in the nano and micrometer range (down to tens of nm; (Shin 2007)). This approach paved the way to studies investigating the role of topographical features of controlled size, geometry, anisotropy and disorder in controlling the migratory behavior of fibroblasts, osteoblast, neurons and several other cell types including stem cells and tumors (Biela et al. 2009; Bettinger et al. 2009; Dalby 2005; Teixeira et al. 2003). Importantly, all these findings relied on fabrication protocols allowing (i) the independent manipulation of the physical parameter under study, (ii) the control over a range of values relevant to cells, and (iii) the possibility to pattern large surface areas (mm^2 to cm^2).

Based on this model, it is possible to describe isolated epithelial cells adhering to a planar substrate. In the absence of physical restrictions, the cells will spread and proliferate in the attempt to occupy the entire available space. The low level of interaction between cells supports a sustained motility, which in a fully isotropic environment displays a gas-like mesenchymal diffusion (Fig. 3.2).

When cell density becomes sufficient to cover the entire substrate, and therefore achieve confluency, the migration speed begins to decrease (Garcia et al. 2015). Cells are now physically confined by the neighbors and cell-to-cell junctions are established (St Johnston and Sanson 2011; Dejana 2004). Collective motion, that is the coherent migration of groups of cells in the same direction at the same time, appears (Angelini et al. 2011). In this phase, the monolayer dynamics is reminiscent of a fluid

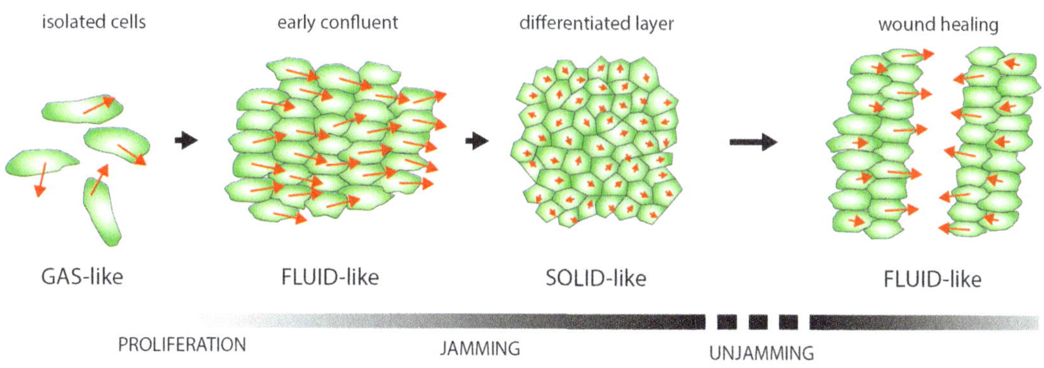

Fig. 3.2 Phases of epithelial migration. Isolated epithelial cells (green) on a flat substrate migrate randomly (the red arrows indicate the amplitude and direction of migration) diffusing like a gas. When contact between neighboring cells is established long-range, fluid-like motion arises. As the monolayer matures and cell density increases cell migration becomes restricted by neighboring cells and motion ceases while the system solidifies. In specific conditions, such as in the presence of a wound or in the emergence of a malignant cancer, collective migration reawakens and enables healing or invasion

(Garrahan 2011). Collective movements decrease with increasing cell density eventually leading to the generation of a mature epithelium where cells assume a characteristic cobblestone shape (St Johnston and Sanson 2011), are stably confined, and largely immotile. This transition to a solid-like state, identified as jamming, is characteristic of mature and functional epithelia (Sadati et al. 2013). In specific conditions, collective cell motility can be re-activated and the system unjammed. This occurs during wound healing or upon invasion of healthy tissues by groups of carcinoma cells (Friedl and Gilmour 2009). The molecular and physical mechanisms governing these complex transitions are not captured by models developed for individual cells. High-throughput genetic approaches have identified the molecular pathways active during collective motility (Fig. 3.2; (Simpson et al. 2008; Vitorino and Meyer 2008)), but even with all this knowledge at hand, several fundamental questions remain to be answered (Park et al. 2016). The main reason for this is the absence of an integrated physical picture enabling the interpretation or even the prediction of experimental results.

In this chapter we will review the most promising experimental models and techniques to study epithelial collectives (Sect. 3.2). Particular attention will be dedicated to traction microscopy and related methods to access the forces involved in the actuation and regulation of emergent collective activities. Based on the available knowledge, the current paradigms of epithelial motility in dense systems will be reviewed (Sect. 3.3). Finally, the endothelium will be treated in a separate section, to highlight the specific differences sustaining the function of this tissue in the regulation of mass transfer (Sect. 3.4).

3.2 Experimental Approaches to Visualize and Study Collective Migration

To shed light on the complexity subtending emergent epithelial phenomena, researchers have developed novel experimental approaches that capture collective cell behaviors (Fig. 3.3). Initial protocols based on live cell imaging of model epithelial monolayers (mostly MDCK cells) cultured of flat substrates to retrieve spatial and temporal maps of structural (i.e. cell shape and number) and kinematic (speed and directionality of cell movements) parameters at different length-scales (Angelini et al. 2011; Petitjean et al. 2010; Zaritsky et al. 2015). Image post-processing generally relied on particle image velocimetry (PIV) for the measurement of velocities and on manual or semi-automated cell segmentation for the extraction of cells shapes and numbers. Therefore, the analysis was strictly dependent on the imaging resolution and was difficult to upscale or parallelize.

The visualization of forces actuating cellular movements added a critical overlapping layer of information to this picture (Fig. 3.4). It was made possible by works which extended classic traction force microscopy (TFM) techniques to whole epithelial sheets (Trepat et al. 2009; Brugues et al. 2014).

Historically, TFM refers to a family of optical methods capturing the deformations induced by cell-generated tractions on compliant substrates (Polacheck and Chen 2016). Due to the importance of cell migration in development, homeostasis and pathology, in the past years, several methods have been developed to estimate cellular forces propelling cell migration.

These methods can be generally divided into continuum or discrete substrate approaches. Continuum substrate methods are based on monitoring the elastic response of a substrate as a continuum. The displacement of any location on the substrate is coupled to the displacements of its neighboring points. Therefore, the entire deformable surface of the substrate acts as force sensor, reporting cellular tractions.

The first successful attempts to visualize cell traction were conducted in reconstituted collagen gels. Fibroblasts were mixed with collagen and casted to obtain a disc. Over time the reduction of the disc diameter was recorded and linked to the cell contractile activity (Bell et al. 1979). This protocol rendered qualitative results averaged on multiple cells within a macroscopic specimen but failed to measure the forces applied by single cells. At the same time an alternative

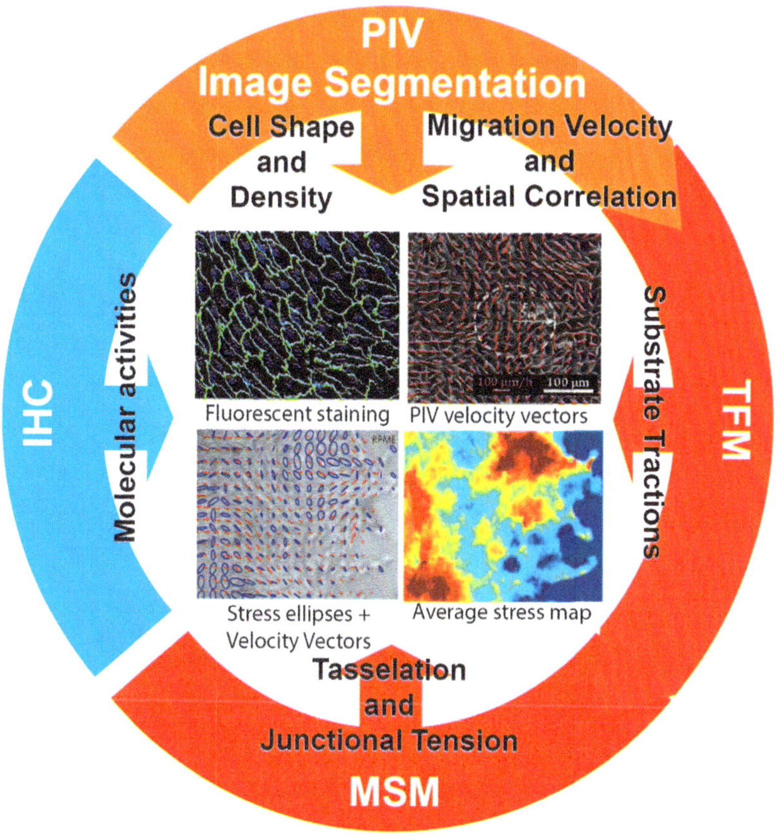

Fig. 3.3 Technologies for the study of collective epithelial motion. Epithelial monolayers grown on flat, transparent substrates are imaged with live-cell microscopy. Particle image velocimetry (PIV) and image segmentation are used to obtain spatially and temporally resolved measures of cell number and morphology, as well as speed and directionality. These parameters are used to obtain values of density and velocity correlation. Traction force microscopy (TFM) uses deformation of compliant substrates to calculate the actuating forces generated by cells through focal adhesions. Monolayer stress microscopy (MSM) uses the same images to derive junctional tensions between cells. All these parameters can be obtained on the same monolayer thus yielding overlapping maps. Immunohistochemistry (IHC) reveals the localization and activation state of molecular elements for which reliable live-cell reporters are not available (e.g. phosphorylated proteins). Current TFM and MSM methods are not compatible with the generation of overlapping IHC maps. (Adapted from Garcia et al. (2015), Tambe et al. (2011), and Trepat et al. (2009))

technique exploited thin silicone films (made of polydimethylsiloxane-PDMS) as substrates for traction force estimation. Chicken fibroblasts were seeded on silicone membranes floating in growth medium (Harris et al. 1980). Cell adhesion and contraction induced visible wrinkling of the PDMS membrane that could be easily captured by bright field microscopy and related to the forces exerted by cells (Fig. 3.5). This approach had several intrinsic limitations. Wrinkles took long to develop and could be larger than cells, resulting in poor spatial and temporal resolution. Additionally, due to the non-linear relationship between forces and wrinkle geometry, a quantitative force analysis proved very complex. Thus, both the wrinkling substrate and the collagen disc assay must be regarded as qualitative approaches unable to provide a reliable force measurement.

The first cell traction assay yielding a reliable quantification of forces was developed by the laboratory of K. Jacobson (Lee et al. 1994).

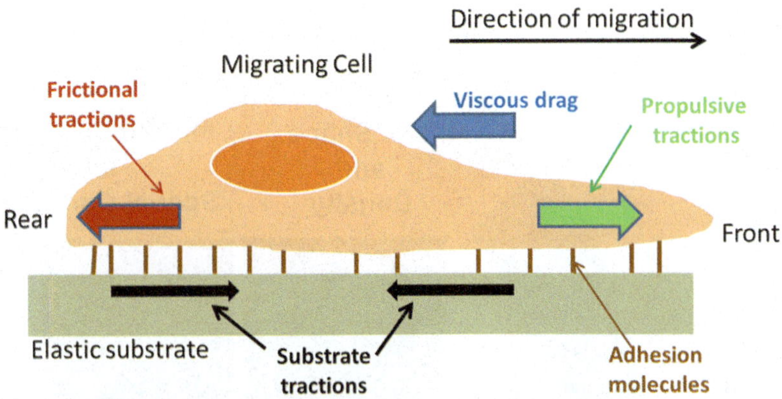

Fig. 3.4 Schematic of a migrating cell and the tractions it generates on the substrate

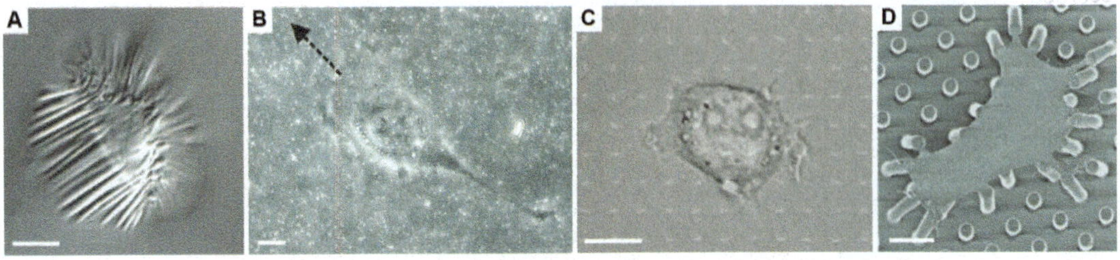

Fig. 3.5 Different principles of traction force assays. (**a**): A keratocyte from gold fish wrinkles a silicone substrate (Burton et al. 1999); (**b**): Fluorescent microspheres embedded in polyacrylamide substrates are used for the detection of substrate surface deformation as a result off forces exerted by a NIH 3 T3 fibroblast (Munevar et al. 2001); (**c**): A NIH 3 T3 fibroblast spread on a regular fluorescent array on PAA (Polio et al. 2012) (**d**): PDMS pillars are deflected by a smooth muscle cell (Tan et al. 2003); Scale bars: 10 μm

The method was based on the visualization of cell-induced substrate displacements through detectable markers introduced in the substrate. The resulting displacement maps were combined with novel quantitative descriptions of the mechanical response of the substrate and resulted in a reasonable estimation of traction forces (Lee et al. 1994; Dembo et al. 1996; Oliver et al. 1995, 1999). To date, this is still a commonly applied method to detect and calculate cell-generated traction forces. The deformable substrates are usually made of silicone or polyacrylamide (PAA) gels while the displacement markers are latex beads (detected in bright field microscopy) or fluorescent microspheres (detected in fluorescent microscopy). The markers are dispersed randomly upon polymerization of the gel (Fig. 3.5b). Cells are then seeded on the surface after specific coating with ECM proteins. The functional principle relies on tracking the displacement of the marker beads using an optical microscopy setup. From these displacements, the traction forces can be calculated, given an accurate continuum model of the substrate based on its mechanical properties and geometry. Studies using displacement-based continuum assays have reported cellular traction stresses in the order of 0.2–10 kPa (Franck et al. 2011; Gutierrez et al. 2011; Munevar et al. 2001; Rape et al. 2011). When combined with 3D imaging techniques, this method is also capable of detecting and measuring out-of-plane deformations, which can arise from both, in-plane and out-of-plane forces (Hur et al. 2009; Legant et al. 2013; Maskarinec et al. 2009).

Among the major drawbacks of this method is the uncertainty on the original position of the beads, which introduces the necessity of a Zero-force reference image to calculate the displacement. To obtain such a reference, the cell(s) under analysis must be removed from the substrate (via enzyme detachment or cell lysis), thereby releasing the applied tension and allowing the markers to return to their original position. While further improvements to this method were introduced with comprehensive continuum theories, 3D finite elements analysis, and more precise detection, the need for a reference image represents an intrinsic limitation for the dynamic study of cellular forces (Butler et al. 2002; Yang et al. 2006).

Reference-free continuum approaches have been developed featuring microcontact printed regular arrays patterning an elastic PAA gel with fluorescent fibronectin regions (Fig. 3.5c), thus creating islands to which cells can adhere. Deformations of the original pattern were recorded and used to extrapolate forces (Polio et al. 2012). However, the spatial resolution of this method is limited by the patterning protocol (lateral spacing: 5 μm). Additionally, cell adhesion was restricted to the fibronectin islands, thus not providing a homogeneous adhesive surface for the cell (similar to the discrete methods described below). A different approach in this direction used microfabrication techniques to pattern fluorescent photoresist on synthetic elastomers (Balaban et al. 2001). While the lateral resolution of this method is improved (2 μm), although still not optimal, the photoresist islands (thickness: 300 nm) create topographic and mechanical inhomogeneity in the substrate which substantially alters the cell activities thus introducing biological artifacts (Curtis and Wilkinson 1997). With the improvement of microfabrication techniques alternative possibilities emerged to tackle the potential drawbacks of continuum methods. Traction forces generated from fibroblasts were measured using arrays of micro machined cantilevers (Galbraith and Sheetz 1998). The deflection of the cantilevers by single focal adhesions allowed the quantitative measurement of adhesion forces at these contacts. Cantilevers are however only able to measure forces in a single point and along one direction and thus cannot assess the full complexity of adhering cells (Wang et al. 2007).

<u>Discrete Methods</u> A powerful approach emerged in the last decade exploiting photoresist and etching techniques to produce arrays of elastic PDMS pillars or posts. The pillar array was used as fakir-bed substrate for cell adhesion by coating the flat pillar-tops with ECM proteins. Upon cell adhesion pillars of given height, diameter and PDMS composition can bend (Fig. 3.5d). Similarly, to the cantilever method described above, the pillar deflection can be optically tracked and used to calculate the forces applied by the cells with straightforward geometrical assumptions (Tan et al. 2003). In this method the spatial resolution is drastically increased and is only limited by the minimal inter-pillar distance that can be reliably fabricated (4–10 μm). This technology provides a reference-free dynamic force sensor array and has been used to demonstrate fundamental aspect of focal adhesion maturation and disassembly. It has e.g. been used to measure the forces exerted per FA in various cell types, ranging from 1 to 10 nN per FA (Fu et al. 2010). In addition to the intrinsic limitation to the detection of planar, 2D forces, the pillar array represents a discontinuous adhesive and a strongly structured (intrusive) substrate, both of which can modify cell behavior as compared to continuum substrates. Therefore, despite providing an elegant and precise method to measure forces, this approach still brings with it considerable inherent limitations and may introduce significant artifacts (Curtis and Wilkinson 1997).

A highly-sensitive, high resolution and reference-free continuum method to measure substrate deformations and render cell-generated traction forces in 3D has been recently developed by our lab (Fig. 3.6). Our method opens the application of TFM to a vast range of biological contexts, ranging from single cell migration to the generation or re-generation of multicellular tissues (Bergert et al. 2016).

In summary, the use of TFM has been instrumental to reveal unexpected force patterns propelling the advancement of epithelial edges during wound healing or other specific epithelial activities (reviewed in the next sections (Trepat et al. 2009, 2010)). In addition, the groups of Prof. Fredberg and Prof. Trepat have recently introduced a computational method to extract values of inter-cellular tension from maps of

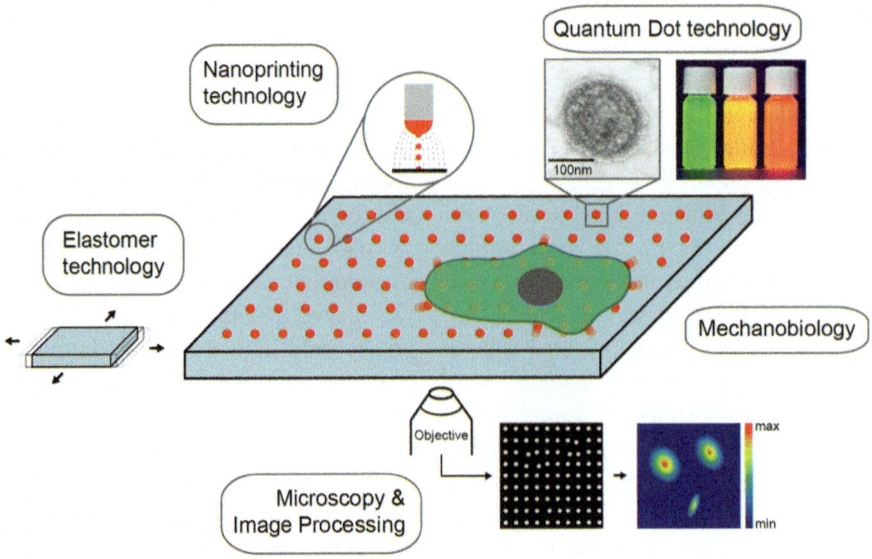

Fig. 3.6 Overview of confocal TFM (cTFM; Bergert et al. 2016)

substrate tractions (Tambe et al. 2011; Kim et al. 2013). Monolayer stress microscopy (MSM) derives the intercellular tensions through a balance of forces within the cell monolayer (Fig. 3.3). This process introduces several simplifying assumptions which limit its applicability to specific experimental conditions such as model wound healing (Tambe et al. 2013). Access to this additional layer of information highlighted the existence of fundamental modes of migration typical of epithelial collectives.

It is now evident that a number of physical and molecular elements must be considered when approaching the study of collective epithelial migration. Typical examples are the jamming and unjamming transitions of epithelia. The generation of an integrative environment where all the independent parameters can be decoupled and integrated within a physical framework shall be readily applied to the study of master molecular regulators for the interpretation of specific physiological or pathological phenomena. This basic knowledge is not restricted to fundamental biology but has already demonstrated its potential in understanding the behavior of epithelial tissues during pathological events such as the response to compressive stresses in asthma (Park et al. 2015). Further important applications include the investigation of the role of genes which are deregulated in cancer patients and that may be implicated in the metastatic invasion of carcinoma (Tam and Weinberg 2013) or on the effect of disturbed hemodynamic conditions in the development of cardiovascular pathologies. The extent to which these cellular activities are relevant for the *in vivo* development or progression of disease is yet to be determined and will be the subject of great future attention by multidisciplinary studies.

3.3 Intrinsic Types of Collective Epithelial Migration

In the following paragraphs the basic parameters and descriptors which can be used to define typical collective responses in epithelial monolayers will be reviewed. The measurements and evaluation of these variable are useful to distil specific patterns of migration, force generation, or cell shape change which can be used as fingerprints of innate epithelial responses.

Spatially and temporally resolved values of cell speed or root-mean-square (rms) velocity (v_{rms}), and of time dependent velocity-velocity spatial correlation (C_{vv}), angular correlation and correlation length (ξ_{vv}) can be obtained applying a particle image velocimetry (PIV)-based approach to time-lapses of collective cell migration (Milde et al. 2012). These time-resolved images are acquired by means of transmission imaging, on reconstituted epithelial monolayers on flat, transparent substrates.

The workflow applies a set of local and global filtering and registering operations followed by a PIV analysis of the optical flow. These operations enable a robust quantification of cell migration dynamics. Values of cell density ($\boldsymbol{\rho}$) and shape (defined as $\boldsymbol{p}$ = (cell perimeter)/$\sqrt{}$(cell area)) are harvested as a function of space and time using segmentation and tracking algorithms (Tarnawski et al. 2013).

This analysis requires additional fluorescent images of cells expressing a nuclear marker (as for example Histone-2B-GFP, (Panagiotakopoulou et al. 2016)) or stained with a live dye (e.g. SiR-Hoechst; (Lukinavicius et al. 2015)). The resulting integration of mechanical, kinematic, and structural data enables their correlation at various length and time scales. The additional level of information retrieved becomes immediately available to validate physical models which capture the motion of individual cells in a monolayer or of global activities such as cell jamming and unjamming (Bi et al. 2015). Examples include, but are not limited to:

- The physiological principle of plithotaxis foresees the tendency of each individual cell within a fluid-like epithelial collective (Fig. 3.2) to adapt its shape and motion in order to minimize the local intercellular shear stress (Tambe et al. 2011). Detection of this guidance mechanism requires access to correlated values of $\boldsymbol{p}$, junctional tension, and v_{rms}. This analysis is useful in resolving alternative guidance mechanisms that drive collective motion in the presence of external directional stimuli including chemotaxis (guidance by gradients of soluble molecules), durotaxis (by gradients of substrate rigidity; (Sunyer et al. 2016)) and haptotaxis (gradients of adhesion points; (Arnold et al. 2008)).
- In the presence of an open space (such as a wound) epithelial cells tend to align their substrate tractions in the direction of wound closure (Kim et al. 2013). Cells following this principle, named kenotaxis, are revealed by correlated analysis of substrate tractions, $\boldsymbol{p}$, and v_{rms}.
- Self-propelled particle models link the jamming transition to a solid-like phase (Fig. 3.2) with an increase of density (Vicsek et al. 1995). This passage is anticipated by the emergence of collective particle motions in the monolayer. Correlations between experimental measures of $\boldsymbol{\rho}$, C_{vv}, and ξ_{vv} provide a direct validation for these models (Garcia et al. 2015).
- Unjamming phenomena and reactivation of cell streaming in confluent epithelia (i.e. at constant density) are typical of embryonic development (Schotz et al. 2013). These transitions are captured by vertex models encoding single-cell properties such as cells shape and cell-cell adhesion (Bi et al. 2015). Correlation between experimental measures of $\boldsymbol{p}$, substrate tractions, intercellular tension and ξ_{vv} are required for the experimental validation of these models (Park et al. 2016).

3.4 The Strange Case of the Endothelium

Although both the endothelium and epithelium act as barriers between different compartments (Kinne 1997), there are major differences between them, including the location, structure, and functions of the two cell types forming these interfaces (Dejana 2004; Kinne 1997). Endothelial cells that form the endothelium in a single thin layer of cells, lacking the tight packed morphology of epithelial cells. This specific configuration allows the controlled passage of molecules of water and oxygen to access the surrounding tissues. The integrity of endothelial monolayers is fundamental for the homoeostasis of the vascular system and plays an important role in its physiological function (Dejana et al. 2009). Indeed, it provides a functional barrier to retain circulating blood, to control blood-tissue exchanges, to recruit blood cells, and to regulate vascular tone.

Endothelial tissue integrity is ensured by the formation of adhesive structures (endothelial junctions) between adjacent cells (Dejana 2004; Lampugnani et al. 2018). These junctional complexes maintain homeostasis of blood vessels, while retaining their capacity to reorganize during angiogenesis. Both adherens and tight junction complexes, connecting adjacent cells, are characterized by a high degree of plasticity. They can rapidly respond to extracellular environmental changes, such as pro- and anti-angiogenic, and inflammatory stimuli, shear stress and blood flow inducing substantial and reversible alterations of the endothelial barrier functions (Orsenigo et al. 2012). The adherens junction protein Vascular Endothelial (VE)-cadherin is an endothelial-specific adhesion molecule located at junctions of endothelial cells and responsible for barrier architecture and function (Lampugnani et al. 2018). VE-cadherin is a transmembrane protein that promotes homophilic interactions, forming a pericellular zipper-like structure along cell boundaries. The association of the C-terminus domain of VE-cadherin with cytoplasmic proteins is needed for its adhesive functions (Lampugnani et al. 2018). Through its cytoplasmic tail, VE-cadherin binds both cytoskeletal and signaling proteins, which in turn allow the anchoring of cadherin to the actin microfilaments and the activation of outside-in signaling (Lampugnani et al. 1995). The association with actin is required for the maintenance of cell shape and polarity, cell movement, stabilization of the junctions and the dynamic regulation of their opening and closure (Oldenburg and de Rooij 2014). Hemodynamic forces, including flow-generated wall shear stress (WSS) and wall deformation (WD) create a complex mechanical environment that impacts on the cellular signals and on the stability of cell-to-cell junctions (Gimbrone Jr 1999) thus coordinating individual endothelial cells in the collective control of vascular homeostasis (McCarron et al. 2017). Endothelial junctions need therefore to be structurally and functionally dynamic, by locally rearranging their continuous organization to create and close intercellular gaps (Orsenigo et al. 2012; Dejana et al. 2017).

Endothelial cell motility is occurring during development as well as in many pathophysiologies in the adult life (Krenning et al. 2016; Guerrero and McCarty 2018). Collective motility of interacting cells is the central feature of these phenomena. These movements are required for important processes such as vasculogenesis, angiogenesis, wound healing and revascularization of injured tissues. Upon mechanical damage, endothelial collective migration is critical to restore an intact monolayer to maintain the functions of the vasculature. Collective migration is characterized by cohesive group of cells that remain physically and functionally connected through cadherin-based junctions coordinating their actin dynamics and intracellular signaling (Yang et al. 2016; Wimmer et al. 2012). This multicellular structural and functional unit is polarized, generates traction and protrusion forces, and deposits and remodels extracellular matrix along the migration route (Friedl and Gilmour 2009). In this multicellular context, cells coordinate migration with adjacent cells ensuring efficient collective movement (Vitorino and Meyer 2008; Rupp et al. 2004; Vitorino et al. 2011). To synchronize this dynamic equilibrium, cells use signaling systems that integrate different responses from growth factor receptors, cell-to-cell junctions, and cell-extracellular matrix interactions (Pignatelli 1998; Gupton and Waterman-Storer 2006; Ogita and Takai 2008). Collective cell migration requires the establishment of a hierarchy of cellular identities that coordinates the movement of the cells (Fig. 3.7).

During sprouting angiogenesis, which occurs in physiological and pathological conditions, or wound healing (Gerhardt 2008), a single endothelial leader cell (tip cell) guides nascent vessels and is followed by other endothelial cells (stalk cells) (Friedl and Gilmour 2009; Herbert and Stainier 2011). The tip cell is highly motile, it is localized at the protrusion of growing vessels and displays an aggressive phenotype characterized by marked stress fibers, enlarged focal contacts and ruffling lamellipodia that translate guidance cues into directional migration (Phng et al.

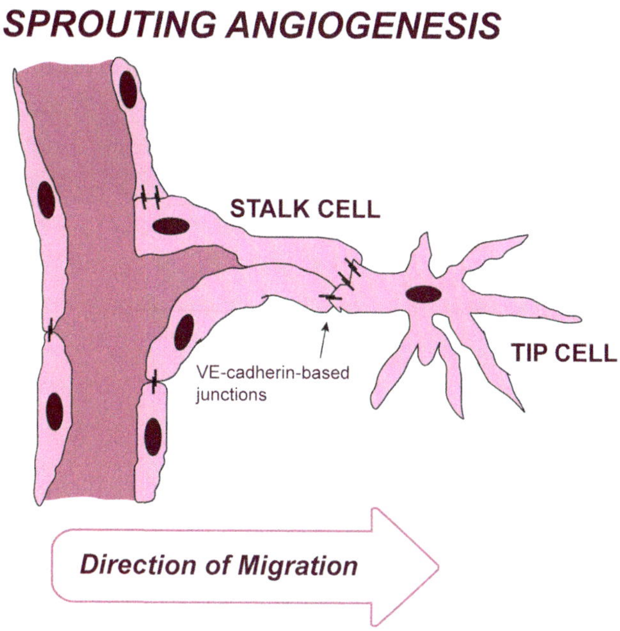

Fig. 3.7 Schematic of the different cellular identities that characterize sprouting angiogenesis

2013). The stress fibers physically link endothelial cells together through the adherens junction complexes to transmit mechanical forces.

The tip cells is then followed by the stalk cells, characterized by proliferative activity, which elongate the vessel sprout to form the main trunk of the developing vessel (Jacobs and Gavard 2018). How this hierarchical organization is established and sustained is one of the key factor to understand the process of collective cell migration. Tip and stalk differentiation is driven and regulated by VEGF (vascular endothelial growth factor) and Notch signaling pathways. In particular, extracellular gradient of VEGF activates VEGF receptor signaling in tip cells that, in turn, induces the expression of Notch ligand Delta-like 4. This, in turn, activates Notch signaling in stalk cells, and in parallel suppresses VEGF signaling in stalk cells to prevent tip cell behavior (Herbert and Stainier 2011). However, tip and stalk cell fates are not permanently defined. Indeed, there is a competition between endothelial cells that causes cell position exchange and triggers the dynamic switch between tip and stalk cell phenotypes both in vitro and in vivo (Jakobsson et al. 2010). Thus, the directionality of the guided sprouting process is not succeeded at the single cellular level of a tip cell, but is the overall result of a collective behavior. Due to the complexity of the range of motions coupled between endothelial cells which results in the overall collective migration, this process is still not completely understood and a better knowledge would be essential for the development of novel therapeutics and tissue engineering approaches that can be used to treat endothelial dysfunction and vascular diseases.

References

Angelini TE, Hannezo E, Trepat X, Marquez M, Fredberg JJ, Weitz DA (2011) Glass-like dynamics of collective cell migration. Proc Natl Acad Sci U S A 108:4714–4719

Arboleda-Estudillo Y, Krieg M, Stuhmer J, Licata NA, Muller DJ, Heisenberg CP (2010) Movement directionality in collective migration of germ layer progenitors. Curr Biol 20:161–169

Arnold M, Hirschfeld-Warneken VC, Lohmuller T, Heil P, Blummel J, Cavalcanti-Adam EA et al (2008) Induction of cell polarization and migration by a gradient of nanoscale variations in adhesive ligand spacing. Nano Lett 8:2063–2069

Arnold M, Schwieder M, Blummel J, Cavalcanti-Adam EA, Lopez-Garcia M, Kessler H et al (2009) Cell interactions with hierarchically structured nano-patterned adhesive surfaces. Soft Matter 5:72–77

Balaban NQ, Schwarz US, Riveline D, Goichberg P, Tzur G, Sabanay I et al (2001) Force and focal adhesion assembly: a close relationship studied using elastic micropatterned substrates. Nat Cell Biol 3:466–472
Bell E, Ivarsson B, Merrill C (1979) Production of a tissue-like structure by contraction of collagen lattices by human fibroblasts of different proliferative potential in vitro. Proc Natl Acad Sci U S A 76:1274–1278
Bergert M, Lendenmann T, Zundel M, Ehret AE, Panozzo D, Richner P et al (2016) Confocal reference free traction force microscopy. Nat Commun 7:12814
Bettinger CJ, Langer R, Borenstein JT (2009) Engineering substrate topography at the micro- and nanoscale to control cell function. Angew Chem Int Ed Engl 48:5406–5415
Bi DP, Lopez JH, Schwarz JM, Manning ML (2015) A density-independent rigidity transition in biological tissues. Nat Phys 11:1074–1079
Biela SA, Su Y, Spatz JP, Kemkemer R (2009) Different sensitivity of human endothelial cells, smooth muscle cells and fibroblasts to topography in the nano-micro range. Acta Biomater 5:2460–2466
Brugues A, Anon E, Conte V, Veldhuis JH, Gupta M, Colombelli J et al (2014) Forces driving epithelial wound healing. Nat Phys 10:683–690
Burton K, Park JH, Taylor DL (1999) Keratocytes generate traction forces in two phases. Mol Biol Cell 10:3745–3769
Butler JP, Tolic-Norrelykke IM, Fabry B, Fredberg JJ (2002) Traction fields, moments, and strain energy that cells exert on their surroundings. Am J Physiol Cell Physiol 282:C595–C605
Cavalcanti-Adam EA, Volberg T, Micoulet A, Kessler H, Geiger B, Spatz JP (2007) Cell spreading and focal adhesion dynamics are regulated by spacing of integrin ligands. Biophys J 92:2964–2974
Curtis A, Wilkinson C (1997) Topographical control of cells. Biomaterials 18:1573–1583
Dalby MJ (2005) Topographically induced direct cell mechanotransduction. Med Eng Phys 27:730–742
Dejana E (2004) Endothelial cell-cell junctions: happy together. Nat Rev Mol Cell Biol 5:261–270
Dejana E, Tournier-Lasserve E, Weinstein BM (2009) The control of vascular integrity by endothelial cell junctions: molecular basis and pathological implications. Dev Cell 16:209–221
Dejana E, Hirschi KK, Simons M (2017) The molecular basis of endothelial cell plasticity. Nat Commun 8:14361
Dembo M, Oliver T, Ishihara A, Jacobson K (1996) Imaging the traction stresses exerted by locomoting cells with the elastic substratum method. Biophys J 70:2008–2022
Ferrari A, Cecchini M (2011) Cells on patterns. In: del Campo A, Arzt E (eds) Generating micro- and nanopatterns on polymeric materials. Wiley-VCH Verlag GmbH & Co. KGaA, Weinheim
Franck C, Maskarinec SA, Tirrell DA, Ravichandran G (2011) Three-dimensional traction force microscopy: a new tool for quantifying cell-matrix interactions. PLoS One 6:e17833
Friedl P, Gilmour D (2009) Collective cell migration in morphogenesis, regeneration and cancer. Nat Rev Mol Cell Biol 10:445–457
Fu J, Wang YK, Yang MT, Desai RA, Yu X, Liu Z et al (2010) Mechanical regulation of cell function with geometrically modulated elastomeric substrates. Nat Methods 7:733–736
Galbraith CG, Sheetz MP (1998) Forces on adhesive contacts affect cell function. Curr Opin Cell Biol 10:566–571
Garcia S, Hannezo E, Elgeti J, Joanny JF, Silberzan P, Gov NS (2015) Physics of active jamming during collective cellular motion in a monolayer. Proc Natl Acad Sci U S A 112:15314–15319
Garrahan JP (2011) Dynamic heterogeneity comes to life. Proc Natl Acad Sci U S A 108:4701–4702
Geiger B, Spatz JP, Bershadsky AD (2009) Environmental sensing through focal adhesions. Nat Rev Mol Cell Biol 10:21–33
Gerhardt H (2008) VEGF and endothelial guidance in angiogenic sprouting. Organogenesis 4:241–246
Gimbrone MA Jr (1999) Vascular endothelium, hemodynamic forces, and atherogenesis. Am J Pathol 155:1–5
Glass R, Moller M, Spatz JP (2003) Block copolymer micelle nanolithography. Nanotechnology 14:1153–1160
Guerrero PA, McCarty JH (2018) Integrins in vascular development and pathology. Adv Pharmacol 81:129–153
Gupton SL, Waterman-Storer CM (2006) Spatiotemporal feedback between actomyosin and focal-adhesion systems optimizes rapid cell migration. Cell 125:1361–1374
Gutierrez E, Tkachenko E, Besser A, Sundd P, Ley K, Danuser G et al (2011) High refractive index silicone gels for simultaneous total internal reflection fluorescence and traction force microscopy of adherent cells. PLoS One 6:e23807
Harris AK, Wild P, Stopak D (1980) Silicone rubber substrata: a new wrinkle in the study of cell locomotion. Science 208:177–179
Herbert SP, Stainier DY (2011) Molecular control of endothelial cell behaviour during blood vessel morphogenesis. Nat Rev Mol Cell Biol 12:551–564
Holmes SJ (1914) The behavior of the epidermis of amphibians when cultivated outside the body. J Exp Zool 17:281–295
Hur SS, Zhao Y, Li YS, Botvinick E, Chien S (2009) Live cells exert 3-dimensional traction forces on their substrata. Cell Mol Bioeng 2:425–436
Hynes RO (1992) Integrins: versatility, modulation, and signaling in cell adhesion. Cell 69:11–25
Jacobs KA, Gavard J (2018) 3D endothelial cell migration. Methods Mol Biol 1749:51–58
Jakobsson L, Franco CA, Bentley K, Collins RT, Ponsioen B, Aspalter IM et al (2010) Endothelial cells dynamically compete for the tip cell position during angiogenic sprouting. Nat Cell Biol 12:943–953

Ji L, Lim J, Danuser G (2008) Fluctuations of intracellular forces during cell protrusion. Nat Cell Biol 10:1393–1400
Keller R (2012) Developmental biology. Physical biology returns to morphogenesis. Science 338:201–203
Kim JH, Serra-Picamal X, Tambe DT, Zhou EH, Park CY, Sadati M et al (2013) Propulsion and navigation within the advancing monolayer sheet. Nat Mater 12:856–863
Kinne RK (1997) Endothelial and epithelial cells: general principles of selective vectorial transport. Int J Microcirc Clin Exp 17:223–230
Krenning G, Barauna VG, Krieger JE, Harmsen MC, Moonen JR (2016) Endothelial plasticity: shifting phenotypes through force feedback. Stem Cells Int 2016:9762959
Lammermann T, Sixt M (2009) Mechanical modes of 'amoeboid' cell migration. Curr Opin Cell Biol 21:636–644
Lampugnani MG, Corada M, Caveda L, Breviario F, Ayalon O, Geiger B et al (1995) The molecular organization of endothelial cell to cell junctions: differential association of plakoglobin, beta-catenin, and alpha-catenin with vascular endothelial cadherin (VE-cadherin). J Cell Biol 129:203–217
Lampugnani MG, Dejana E, Giampietro C (2018) Vascular endothelial (VE)-cadherin, endothelial adherens junctions, and vascular disease. Cold Spring Harb Perspect Biol 10:a029322
Lauffenburger DA, Horwitz AF (1996) Cell migration: a physically integrated molecular process. Cell 84:359–369
Lee J, Leonard M, Oliver T, Ishihara A, Jacobson K (1994) Traction forces generated by locomoting keratocytes. J Cell Biol 127:1957–1964
Legant WR, Choi CK, Miller JS, Shao L, Gao L, Betzig E et al (2013) Multidimensional traction force microscopy reveals out-of-plane rotational moments about focal adhesions. Proc Natl Acad Sci U S A 110:881–886
Lo CM, Wang HB, Dembo M, Wang YL (2000) Cell movement is guided by the rigidity of the substrate. Biophys J 79:144–152
Lukinavicius G, Blaukopf C, Pershagen E, Schena A, Reymond L, Derivery E et al (2015) SiR-hoechst is a far-red DNA stain for live-cell nanoscopy. Nat Commun 6:8497
Maskarinec SA, Franck C, Tirrell DA, Ravichandran G (2009) Quantifying cellular traction forces in three dimensions. Proc Natl Acad Sci U S A 106:22108–22113
McCarron JG, Lee MD, Wilson C (2017) The endothelium solves problems that endothelial cells do not know exist. Trends Pharmacol Sci 38:322–338
Milde F, Franco D, Ferrari A, Kurtcuoglu V, Poulikakos D, Koumoutsakos P (2012) Cell Image Velocimetry (CIV): boosting the automated quantification of cell migration in wound healing assays. Integr Biol 4:1437–1447
Munevar S, Wang Y, Dembo M (2001) Traction force microscopy of migrating normal and H-ras transformed 3T3 fibroblasts. Biophys J 80:1744–1757
Ogita H, Takai Y (2008) Cross-talk among integrin, cadherin, and growth factor receptor: roles of nectin and nectin-like molecule. Int Rev Cytol 265:1–54
Oldenburg J, de Rooij J (2014) Mechanical control of the endothelial barrier. Cell Tissue Res 355:545–555
Oliver T, Dembo M, Jacobson K (1995) Traction forces in locomoting cells. Cell Motil Cytoskeleton 31:225–240
Oliver T, Dembo M, Jacobson K (1999) Separation of propulsive and adhesive traction stresses in locomoting keratocytes. J Cell Biol 145:589–604
Orsenigo F, Giampietro C, Ferrari A, Corada M, Galaup A, Sigismund S et al (2012) Phosphorylation of VE-cadherin is modulated by haemodynamic forces and contributes to the regulation of vascular permeability in vivo. Nat Commun 3:1208
Paluch EK, Aspalter IM, Sixt M (2016) Focal adhesion-independent cell migration. Annu Rev Cell Dev Biol 32:469–490
Panagiotakopoulou M, Bergert M, Taubenberger A, Guck J, Poulikakos D, Ferrari A (2016) A nanoprinted model of interstitial cancer migration reveals a link between cell deformability and proliferation. ACS Nano 10:6437–6448
Park JA, Kim JH, Bi D, Mitchel JA, Qazvini NT, Tantisira K et al (2015) Unjamming and cell shape in the asthmatic airway epithelium. Nat Mater 14:1040–1048
Park JA, Atia L, Mitchel JA, Fredberg JJ, Butler JP (2016) Collective migration and cell jamming in asthma, cancer and development. J Cell Sci 129:3375–3383
Pelham RJ Jr, Wang Y (1997) Cell locomotion and focal adhesions are regulated by substrate flexibility. Proc Natl Acad Sci U S A 94:13661–13665
Petitjean L, Reffay M, Grasland-Mongrain E, Poujade M, Ladoux B, Buguin A et al (2010) Velocity fields in a collectively migrating epithelium. Biophys J 98:1790–1800
Petrie RJ, Yamada KM (2015) Fibroblasts lead the way: a unified view of 3D cell motility. Trends Cell Biol 25:666–674
Phng LK, Stanchi F, Gerhardt H (2013) Filopodia are dispensable for endothelial tip cell guidance. Development 140:4031–4040
Pignatelli M (1998) Integrins, cadherins, and catenins: molecular cross-talk in cancer cells. J Pathol 186:1–2
Polacheck WJ, Chen CS (2016) Measuring cell-generated forces: a guide to the available tools. Nat Methods 13:415–423
Polio SR, Rothenberg KE, Stamenovic D, Smith ML (2012) A micropatterning and image processing approach to simplify measurement of cellular traction forces. Acta Biomater 8:82–88
Rape AD, Guo WH, Wang YL (2011) The regulation of traction force in relation to cell shape and focal adhesions. Biomaterials 32:2043–2051

Ridley AJ, Schwartz MA, Burridge K, Firtel RA, Ginsberg MH, Borisy G et al (2003) Cell migration: integrating signals from front to back. Science 302:1704–1709

Rupp PA, Czirok A, Little CD (2004) alphavbeta3 integrin-dependent endothelial cell dynamics in vivo. Development 131:2887–2897

Sadati M, Qazvini NT, Krishnan R, Park CY, Fredberg JJ (2013) Collective migration and cell jamming. Differentiation 86:121–125

Schotz EM, Lanio M, Talbot JA, Manning ML (2013) Glassy dynamics in three-dimensional embryonic tissues. J R Soc Interface 10:20130726

Shin H (2007) Fabrication methods of an engineered microenvironment for analysis of cell-biomaterial interactions. Biomaterials 28:126–133

Simpson KJ, Selfors LM, Bui J, Reynolds A, Leake D, Khvorova A et al (2008) Identification of genes that regulate epithelial cell migration using an siRNA screening approach. Nat Cell Biol 10:1027–1038

St Johnston D, Sanson B (2011) Epithelial polarity and morphogenesis. Curr Opin Cell Biol 23:540–546

Sunyer R, Conte V, Escribano J, Elosegui-Artola A, Labernadie A, Valon L et al (2016) Collective cell durotaxis emerges from long-range intercellular force transmission. Science 353:1157–1161

Tam WL, Weinberg RA (2013) The epigenetics of epithelial-mesenchymal plasticity in cancer. Nat Med 19:1438–1449

Tambe DT, Hardin CC, Angelini TE, Rajendran K, Park CY, Serra-Picamal X et al (2011) Collective cell guidance by cooperative intercellular forces. Nat Mater 10:469–475

Tambe DT, Croutelle U, Trepat X, Park CY, Kim JH, Millet E et al (2013) Monolayer stress microscopy: limitations, artifacts, and accuracy of recovered intercellular stresses. PLoS One 8:e55172

Tan JL, Tien J, Pirone DM, Gray DS, Bhadriraju K, Chen CS (2003) Cells lying on a bed of microneedles: an approach to isolate mechanical force. Proc Natl Acad Sci U S A 100:1484–1489

Tarnawski W, Kurtcuoglu V, Lorek P, Bodych M, Rotter J, Muszkieta M et al (2013) A robust algorithm for segmenting and tracking clustered cells in time-lapse fluorescent microscopy. IEEE J Biomed Health Inform 17:862–869

Teixeira AI, Abrams GA, Bertics PJ, Murphy CJ, Nealey PF (2003) Epithelial contactguidance on well-defined micro- and nanostructured substrates. J Cell Sci 116:1881–1892

Trepat X, Wasserman MR, Angelini TE, Millet E, Weitz DA, Butler JP et al (2009) Physical forces during collective cell migration. Nat Phys 5:426–430

Trepat X, Fabry B, Fredberg JJ (2010) Pulling it together in three dimensions. Nat Methods 7:963–965

Tzvetkova-Chevolleau T, Stephanou A, Fuard D, Ohayon J, Schiavone P, Tracqui P (2008) The motility of normal and cancer cells in response to the combined influence of the substrate rigidity and anisotropic microstructure. Biomaterials 29:1541–1551

Vicsek T, Czirok A, Ben-Jacob E, Cohen II, Shochet O (1995) Novel type of phase transition in a system of self-driven particles. Phys Rev Lett 75:1226–1229

Vitorino P, Meyer T (2008) Modular control of endothelial sheet migration. Genes Dev 22:3268–3281

Vitorino P, Hammer M, Kim J, Meyer T (2011) A steering model of endothelial sheet migration recapitulates monolayer integrity and directed collective migration. Mol Cell Biol 31:342–350

Wang JHC, Lin JS, Yang ZC (2007) Cell traction force microscopy. In: Advanced bioimaging technologies in assessment of the quality of bone and scaffold materials: techniques and applications. Springer, Berlin, pp 227–235

Wimmer R, Cseh B, Maier B, Scherrer K, Baccarini M (2012) Angiogenic sprouting requires the fine tuning of endothelial cell cohesion by the Raf-1/Rok-alpha complex. Dev Cell 22:158–171

Wozniak MA, Modzelewska K, Kwong L, Keely PJ (2004) Focal adhesion regulation of cell behavior. Biochim Biophys Acta 1692:103–119

Yang Z, Lin JS, Chen J, Wang JH (2006) Determining substrate displacement and cell traction fields – a new approach. J Theor Biol 242:607–616

Yang Y, Jamilpour N, Yao B, Dean ZS, Riahi R, Wong PK (2016) Probing leader cells in endothelial collective migration by plasma lithography geometric confinement. Sci Rep 6:22707

Zaritsky A, Welf ES, Tseng YY, Rabadan MA, Serra-Picamal X, Trepat X et al (2015) Seeds of locally aligned motion and stress coordinate a collective cell migration. Biophys J 109:2492–2500

Zeng D, Ferrari A, Ulmer J, Veligodskiy A, Fischer P, Spatz J et al (2006) Three-dimensional modeling of mechanical forces in the extracellular matrix during epithelial lumen formation. Biophys J 90:4380–4391

Continuum Models of Collective Cell Migration

4

Shiladitya Banerjee and M. Cristina Marchetti

Abstract

Collective cell migration plays a central role in tissue development, morphogenesis, wound repair and cancer progression. With the growing realization that physical forces mediate cell motility in development and physiology, a key biological question is how cells integrate molecular activities for force generation on multicellular scales. In this review we discuss recent advances in modeling collective cell migration using quantitative tools and approaches rooted in soft matter physics. We focus on theoretical models of cell aggregates as continuous active media, where the feedback between mechanical forces and regulatory biochemistry gives rise to rich collective dynamical behavior. This class of models provides a powerful predictive framework for the physiological dynamics that underlies many developmental processes, where cells need to collectively migrate like a viscous fluid to reach a target region, and then stiffen to support mechanical stresses and maintain tissue cohesion.

S. Banerjee (✉)
University College London, London, UK
e-mail: shiladitya.banerjee@ucl.ac.uk

M. C. Marchetti
University of California Santa Barbara, Santa Barbara, CA, USA
e-mail: cmarchetti@ucsb.edu

Keywords

Continuum modelling · Cell migration · Cell mechanics · Tissue mechanics · Active matter

4.1 Introduction

In many physiological and developmental contexts, groups of cells coordinate their behavior to organize in coherent structures or migrate collectively (Friedl and Gilmour 2009). Many experimental studies have established that these multicellular processes are regulated by the cross-talk between cell-cell adhesions, cell interaction with the extracellular matrix, and myosin-based contractility of the cell cortex (Ladoux and Mège 2017). Importantly, faithful execution of multicellular processes requires both biochemical signaling and mechanical force transmission.

A well-studied multicellular process is wound healing, where epithelial cells march in unison to fill in a gap in the tissue (Begnaud et al. 2016; Fenteany et al. 2000). Although the cells at the front of the advancing monolayer often show large, spread-out lamellipodia and an almost mesenchymal phenotype, long-range collective migration is not simply achieved via the pulling action of such *leader* cells on a sheet of inert followers (Trepat et al. 2009). In fact, traction forces transmitted to the extracellular

C. A. M. La Porta, S. Zapperi (eds.), *Cell Migrations: Causes and Functions*, Advances in Experimental Medicine and Biology 1146,
https://doi.org/10.1007/978-3-030-17593-1_4

matrix are found to remain significant well behind the leading edge of the tissue, indicating that cells in the bulk participate in force generation and transmission. This observation, together with the presence of spread-out cells with large cryptic lamellipodia throughout the monolayer (Farooqui and Fenteany 2005), indicates that, although leader cells at the sheet edge provide guidance for migration, they do not play a unique role in force generation. Instead, a new paradigm has emerged where collective migration is associated with long-range forces extending throughout the tissue, with waves of propagating mechanical stress that are sustained by biochemical signaling at the molecular scale (Serra-Picamal et al. 2012; Trepat et al. 2009). These waves of stress and cellular deformation provide a mechanism for information transmission, much like sound in air. Such mechanical waves have been shown to drive periodic cycles of effective stiffening and fluidification in expanding cell monolayers (Serra-Picamal et al. 2012) and coherent vortical or standing motions in confined ones (Deforet et al. 2014; Doxzen et al. 2013; Notbohm et al. 2016).

Multicellularity and collective migration is intimately related to the materials properties of tissues – viscoelastic materials with both fluid and solid-like behavior. In morphogenesis, for instance, cells must sort and flow like a liquid to reach the right location, but then stiffen and support mechanical stresses once the tissue has achieved the desired structure (Lecuit et al. 2011). Recent experiments have suggested that dense tissues may be in a *glassy* or *jammed* state, where local cell rearrangements are rare and energetically costly. A relatively small change in tissue mechanical parameters may trigger a change from an elastic response to viscous fluid-like behavior, where individual cells are highly motile and rearrange continuously (Angelini et al. 2010, 2011). Indeed living tissues appear to have well-defined mechanical properties, some familiar from conventional matter, such as elastic moduli (Discher et al. 2005) and surface tension (Foty et al. 1994), others unique to living systems, such as *homeostatic pressure*, proposed theoretically as a factor controlling tumor growth (Shraiman 2005; Basan et al. 2009).

Just like intermolecular forces yield the emergence of materials properties in nonliving matter, cell-cell interactions, mediated by cadherins, play a crucial role in controlling the macroscopic properties of groups of cells and tissues (Maruthamuthu et al. 2011; Mertz et al. 2013). The collective mechanics of living matter, however, is more complex than that of inert materials as individual cell activity competes with cell-cell interactions in controlling the large scale behavior of cell assemblies. In addition, physical models of collective cell behavior must also incorporate interactions of cells with the extracellular matrix. In other words, the coupling of cells to their surroundings is affected by intracellular contractility and cell-cell interactions, which in turn can be actively regulated by the environment, in a complex feedback loop unique to living matter. Finally, unlike inert materials where phase changes are controlled by externally tuning parameters such as temperature and density, living matter can tune itself between states with different macroscopic properties through the regulation of molecular scale and genetic processes that drive motility, division, death and phenotypical changes. A quantitative understanding of the relative importance of mechanical and biochemical mechanisms in controlling the collective tissue properties is beginning to emerge through developments in molecular biology, microscopy, super-resolution imaging and force measurement techniques (Roca-Cusachs et al. 2017). These advances provide an ideal platform for constructing quantitative physical models that account for the role of active cellular processes in controlling collective mechanics of motile and deformable multicellular structures.

Theoretical modeling of multicellular processes can be divided broadly into two classes. The first encompasses discrete mesoscale models that incorporate some minimal features of individual cells, such as contractility and motility, and then examine how cell-cell interactions and coupling to the environment determine materials properties at the tissue scale. This class includes models of cells as active particles endowed with persistent motility (Basan et al. 2013; Camley and Rappel 2017), as well as models that have been used extensively in developmental biology, such as Vertex (Fletcher et al. 2014; Honda and Eguchi 1980), Voronoi (Bi et al. 2016; Li and Sun 2014) and Cellular Potts models (Graner and Glazier 1992) that are designed to capture the behavior of confluent tissues, where there are no gaps nor overlaps between cells. Vertex and Voronoi models describe cells as irregular polygons tiling the plane and are defined by an energy functional that tends to adjust the area and perimeter of each cell to target values (Farhadifar et al. 2007). Recent modifications have also endowed these mesoscopic models with cell motility (Barton et al. 2017; Bi et al. 2016; Staddon et al. 2018) and active contractility (Noll et al. 2017). Vertex models have been employed successfully to quantify how intercellular forces control shape at both the cell and tissue scale under the assumption of force balance at every vertex of the cellular network (Farhadifar et al. 2007). An active version of the Voronoi model was recently shown to exhibit a liquid-solid transition of confluent epithelia tuned by motility and cell shape, which in turns encodes information about the interplay between cortex contractility and cell-cell adhesion (Bi et al. 2016). An intriguing prediction of this work is that individual cell shape, that can be inferred directly from cell imaging segmentation, provides a measure of tissue rigidity (Bi et al. 2015).

The second class of theoretical work encompasses continuum models, such as phase field (Ziebert et al. 2011) and active gel models (Prost et al. 2015), where a cell sheet is described as a fluid or an elastic continuum, with couplings to internal degrees of freedom that account for active processes, such as contractility and cellular polarization. Continuum models have been shown to account for the heterogeneous spatial distribution of cellular stresses inferred from Traction Force Microscopy (Style et al. 2014) in both expanding (Banerjee and Marchetti 2011a; Blanch-Mercader et al. 2017; Serra-Picamal et al. 2012; Trepat et al. 2009) and confined monolayers (Notbohm et al. 2016), and even at the level of individual cells (Oakes et al. 2014). They also capture the mechanical waves observed in these systems (Banerjee et al. 2015; Serra-Picamal et al. 2012). This review does not aim to be comprehensive, and will focus on models of tissue as active continuous media, with an emphasis on models that describe tissue as active *elastic* continua. This class of mechanochemical models has had a number of successes in capturing the tissue scale behavior in adherent (Mertz et al. 2012), confined (Notbohm et al. 2016) and expanding epithelia (Banerjee et al. 2015).

Both the mesoscale and continuum approaches do not attempt to faithfully incorporate intracellular processes, but rather aim at characterizing quantitatively the modes of organization and the materials properties of cell collectives in terms of a few macroscopic parameters, such as cell density and shape, cell-cell adhesiveness, contractility, polarization and division/death rates. Each of these quantities may describe the combined effect of a number of molecular processes and signaling pathways. This approach, inspired from condensed matter physics (Marchetti et al. 2013), aims at providing experimentalists with testable predictions that may allow to correlate classes of signaling pathways to tissue scale organization.

The review is organized as follows. In Sect. 4.2 we describe a dynamical model of cell collectives as active viscoelastic media, coupled to the dynamics of active intracellular processes such as actomyosin contractility and cell polarization. An important aspect of the model is a dynamic feedback between mechanical stresses and regulatory biochemistry which gives rise to rich collective behavior. In Sect. 4.3 we discuss applications of this class of continuum models to describing force transmission in epithelial monolayers, waves in expanding cell sheets, collective cell migration in confinement and during epithelial gap closure. We then compare the quantitative predictions of viscoelastic solid models with fluid models of tissues in Sect. 4.4, describing their equivalence as well as highlighting the key differences. We conclude with a critical discussion of the continuum model limitations and highlight open theoretical questions in understanding the collective behavior of multicellular assemblies (Sect. 4.5).

4.2 Cells as Active Continuous Media

We begin by considering the mechanics of a monolayer of epithelial cells, migrating on a soft elastic matrix (Fig. 4.1a–b), with an average height h much thinner than in-plane cell dimensions (Banerjee and Marchetti 2011a, 2012; Schwarz and Safran 2013). In mechanical equilibrium, the condition of local force-balance translates to $\partial_\beta \Sigma_{\alpha\beta} = 0$, where $\boldsymbol{\Sigma}$ is the three-dimensional stress tensor of the monolayer, with greek indices taking values x, y and z. In-plane force balance is given by

$$\partial_j \Sigma_{ij} + \partial_z \Sigma_{iz} = 0, \tag{4.1}$$

with i, j denoting in-plane coordinates. For a thin cell monolayer we average the cellular force-balance equation over the cell thickness h. We assume that the top surface of the cell is stress free, $\Sigma_{iz}(\mathbf{r}_\perp, z = h) = 0$, whereas at the cell-substrate interface, $z = 0$, the cells experience lateral traction stresses given by $\Sigma_{iz}(\mathbf{r}_\perp, z=0) = T_i(\mathbf{r}_\perp)$. A representative traction stress map for a monolayer expanding in free space is reproduced in Fig. 4.1b, which shows appreciable traction stress penentration throughout the bulk of the tissue. The thickness-averaged force balance equation then reads,

$$h\partial_j \sigma_{ij} = T_i, \tag{4.2}$$

where $\sigma_{ij}(\mathbf{r}_\perp) = \int_0^h (dz/h)\Sigma_{ij}(\mathbf{r}_\perp, z)$ is the in-plane monolayer stress. The force-balance diagram is illustrated in Fig. 4.1c. It is worthwhile to mention that the assumption of in-plane traction forces is a good approximation for fully spread cells making almost zero contact angle with the substrate. During the early stages of spreading and migration, cells can exert appreciable out-of-plane traction forces via rotation of focal adhe-

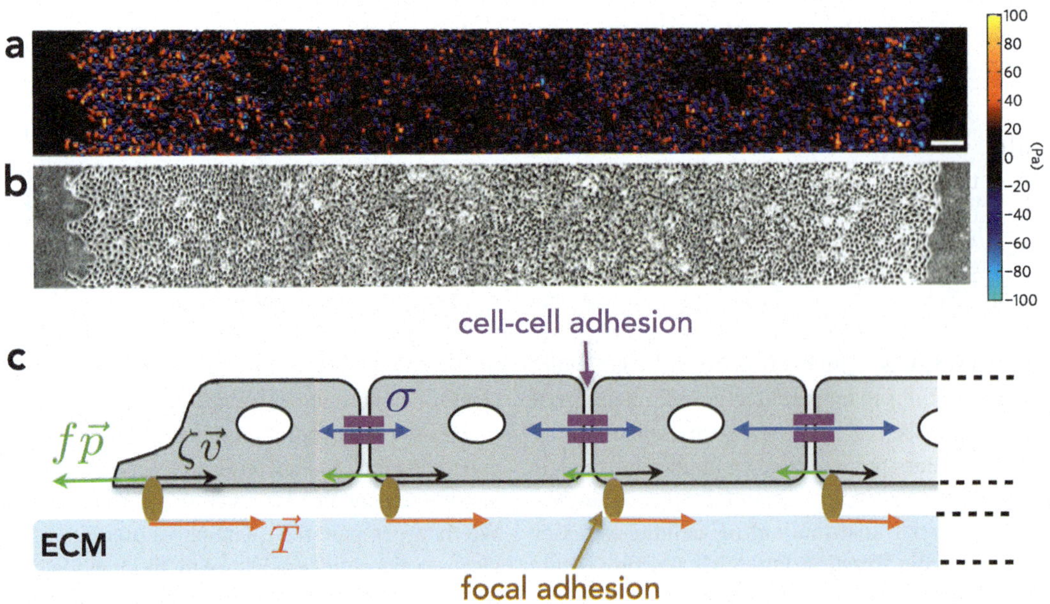

Fig. 4.1 Forces driving collective cell motion. (**a–b**) Radial component of traction stress (**a**) and phase contrast images of an expanding MDCK cell monolayer. (Reproduced from Trepat et al. 2009) (scale bar = 200 μm). (**b**) Schematic of the physical forces acting on the cell monolayer (Notbohm et al. 2016). Tractions exerted by the monolayer on the substrate (ECM) point inward (red arrows) at the monolayer edge and balance the forces due to viscous friction, $\zeta \mathbf{v}$ (black arrows), and polarized motility, $f\mathbf{p}$ (green arrows). The tractions are locally balanced by the divergence of the monolayer stress, $\mathbf{T} = h\nabla \cdot \boldsymbol{\sigma}$

sions (Legant et al. 2013). The quantity T_i is a stress in three dimensions, i.e., a force per unit area. It describes the in-plane traction force per unit area that the cells exert on the substrate. The force-balance equation is supplemented by the mass balance equation, such that the cell density, $\rho(\mathbf{r}_\perp, t)$, obeys the following conservation equation,

$$\partial_t \rho + \nabla \cdot (\rho \mathbf{v}) = \chi \rho, \qquad (4.3)$$

where $\mathbf{v}$ is the velocity field, and χ is the rate of variation in cell density due to cell division or death (Bove et al. 2017). In the following, we assume $\chi = 0$. See Murray and Oster (1984), Ranft et al. (2010), and Yabunaka and Marcq (2017) for continuum models for tissues with explicit consideration of cell division and death.

4.2.1 Constitutive Model for Intercellular Stress

The in-plane cellular stress, $\boldsymbol{\sigma}$, can be decomposed as the sum of intercellular stress, $\boldsymbol{\sigma}^c$, and active stress, $\boldsymbol{\sigma}^a$, originating from active intracellular processes (Fig. 4.2). The form of the constitutive relation for the intercellular stress has been highly debated, given the complex rheology of cellular aggregates (Khalilgharibi et al. 2016). On the timescale of seconds to minutes, living tissues behave elastically, recovering their original shape after a transient application of force (Guevorkian et al. 2010; Phillips and Steinberg 1978). On longer timescales (tens of minutes to hours), cellular aggregates exhibit fluid-like behavior that can arise from cell-cell adhesion turnover, cellular rearrangements, cell division or death (Guillot and Lecuit 2013; Heisenberg and Bellaïche 2013; Ranft et al. 2010). It is therefore commonly assumed that intercellular stresses obey Maxwell visco-elastic constitutive law (Lee and Wolgemuth 2011), described by solid-like response at short time scales and fluid-like behavior at longer time scales.

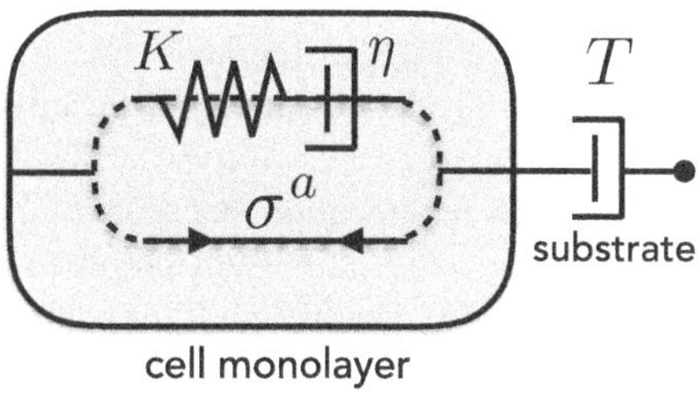

Fig. 4.2 Constitutive elements of the continuum model for collective migration. The viscoelastic and active elements exert stresses in parallel. A local gradient in stress is balanced by the traction exerted by the cell on the substrate, which is modelled by a viscous element

Experimental and computational studies by many groups have shown, however, that stresses imposed on tissues cannot be completely dissipated, and cells support some part of applied tension (Gonzalez-Rodriguez et al. 2013; Harris et al. 2012; Wayne Brodland and Wiebe 2004). In fact rheological experiments have demonstrated that stress relaxation in epithelial monolayers can be described by a spring connected in parallel to a viscous dashpot (Harris et al. 2012; ?). Others have shown that mechanical stress buildup in monolayers occurs in unison with strain accumulation (Serra-Picamal et al. 2012), which can be described by an elastic constitutive law (Mertz et al. 2012; Tambe et al. 2011). Therefore, to describe the dynamic mechanical behavior of cohesive cellular aggregates we assume linear Kelvin-Voigt rheology (Fig. 4.2) (Murray and Oster 1984)

$$\boldsymbol{\sigma}^c = (1 + \tau \partial_t)\big[K \nabla \cdot \mathbf{u}\, \mathbb{1} + \mu \left(\nabla \mathbf{u} + (\nabla \mathbf{u})^T - \nabla \cdot \mathbf{u}\, \mathbb{1}\right)\big], \qquad (4.4)$$

where $\mathbb{1}$ is the identity matrix, $\mathbf{u}$ is the cellular displacement field, K is the compressional elastic modulus, μ is the shear modulus, and τ is the viscoelastic relaxation timescale. The assumption of isotropic elasticity is consistent with stress measurement in cell monolayers using monolayer stress microscopy (Notbohm et al. 2016; Tambe et al. 2011). For simplicity, we have ignored nonlinear contributions to the constitutive relation in Eq. (4.20), which may be essential for stabilizing the dynamical response of living tissues to large mechanical strain (Banerjee et al. 2011, 2017; Köpf and Pismen 2013). In Sect. 4.4, we discuss the quantitative comparisons between elastic and fluid models of tissue rheology. We note that recent experimental studies show evidence for more complex rheological properties, including combinations of active elastic and dissipative response at moderate stretching (?), as well as superelastic behavior at extreme stretching (Latorre et al. 2018).

4.2.2 Active Intracellular Stress

The active intracellular stress stems from contractile forces generated in the actomyosin cytoskeleton in the cell cortex (Murrell et al. 2015), and from actin treadmiling driven by the assembly and disassembly actin filaments. Active contractile stresses depend on the concentration of active actomyosin units, $c(t)$, with the form

$$\boldsymbol{\sigma}^a = \sigma_0(c)\mathbb{1} + \sigma_{an}(c)\mathbf{pp}, \qquad (4.5)$$

where we have introduced the cell polarization or polarity vector, $\mathbf{p}$, which is an internal state variable that controls the local direction of cell motion (Fig. 4.3). $\sigma_0(c)$ and $\sigma_{an}(c)$ are the isotropic and anisotropic components of the active stress due to actomyosin contractility. Note that additional active stress terms of the form $\propto \nabla\mathbf{p}$ are allowed by symmetry in this phenomenological model, leading to renormalization of the elastic modulus to leading order (Banerjee and Marchetti 2011a). Several models for the dependence of σ_0 on c have been proposed, including linear (Banerjee and Marchetti 2011b), logarithmic (Banerjee et al. 2015) and saturating behaviour (Bois et al. 2011). Recent in vitro measurements show that contractile strains accumulate cooperatively as a function of myosin density (Linsmeier et al. 2016), indicating that σ_0 could take the general Hill functional form:

$$\sigma_0(c) = \sigma_0 \frac{c^n}{c_*^n + c^n}, \qquad (4.6)$$

where the constant $n > 1$ indicates cooperative behavior beyond a critical concentration c_*, and $\sigma_0 > 0$ is the magnitude of the contractile stress.

Finally, the force balance equation, Eq. 4.2, requires a constitutive equation for the net traction stress transmitted to the substrate. For a layer of motile cells this is chosen of the form (Fig. 4.1c) (Banerjee et al. 2015)

$$\mathbf{T} = \zeta\mathbf{v} - f\mathbf{p}, \qquad (4.7)$$

where $\mathbf{v} = \partial_t\mathbf{u}$, f is the magnitude of the propulsion force, and ζ is an effective friction coefficient that depends on the rate of focal adhesion turnover (Walcott and Sun 2010). This form for traction in Eq. (4.7) results in local misalignment of traction stress and cell velocity, consistent with experimental findings (Brugués et al. 2014; Notbohm et al. 2016). The propulsion force, $f\mathbf{p}$, drives cell crawling, and depends on the concentration of branched actin in the lamellipodia of migrating cells. For simplicity, we assume that there is a steady concentration of polymerized actin that pushes the cell forward. Dynamic models for the competition between branched and contractile actin have been proposed (Lomakin et al. 2015; Suarez and Kovar 2016). A detailed description of such molecular processes lies beyond the scope of this review, but can be easily incorporated within this framework. The resultant force balance equation is then given by (Figs. 4.1c and 4.2),

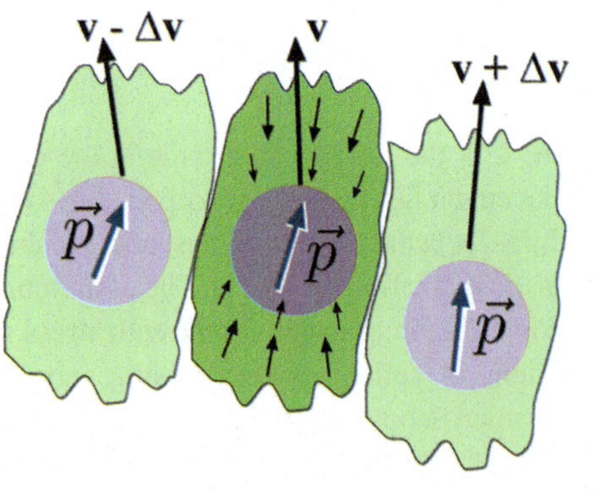

Fig. 4.3 Coordination of cell motion and polarization. Cells align their motion along the polarity vector, **p**, and move with a velocity **v**. Neighboring cells tend to align their polarities, and polarity differences generate a net torque on neighboring cells. Cells also exert a dipole-like contractile stress on the substrate due to actomyosin activity. (Figure adapted from Lee and Wolgemuth 2011)

$$h\nabla \cdot (\boldsymbol{\sigma}^c + \boldsymbol{\sigma}^a) = \zeta \mathbf{v} - f\mathbf{p} + \mathbf{f}_{\text{ext}}, \tag{4.8}$$

where $\mathbf{f}_{\text{ext}}$ is the external force (density) applied to the system. In the absence of external forces or stresses applied at the boundary, the net traction force, when integrated over the entire cell-substrate interface must vanish. This implies a fundamental constraint on the relationship between cell polarity and velocity:

$$\int \mathbf{v} \cdot \mathbf{dA} = \frac{f}{\zeta} \int \mathbf{p} \cdot \mathbf{dA}. \tag{4.9}$$

In the following, we will additionally need to prescribe the dynamics of cell polarization and actomyosin concentration, which regulate active cell motility and the production of contractile stresses.

4.2.3 Mechanochemical Coupling of Cell Motion and Contractility

The dynamics of cell polarization is commonly modeled following the physics of active liquid crystals (Marchetti et al. 2013), a phenomenological approach that requires further justification and scrutiny. The cell polarization vector evolves in time according to,

$$\begin{aligned}\partial_t \mathbf{p} + \beta(\mathbf{p} \cdot \nabla)\mathbf{p} + \mathbf{v} \cdot \nabla \mathbf{v} - \frac{1}{2}(\nabla \times \mathbf{v}) \times \mathbf{v} \\ = a(1 - |\mathbf{p}|^2)\mathbf{p} + \kappa \nabla^2 \mathbf{p} + w\nabla c, \end{aligned} \tag{4.10}$$

where the advective coupling β arises from ATP driven processes such as treadmiling (Ahmadi et al. 2006), the velocity dependent advective terms are borrowed from the nematic liquid crystal literature (Prost 1995), and the Franck elastic constants are both assumed to be equal to κ. Here, a controls the rate of relaxation to a homogeneously polarized cell monolayer, and κ controls the strength of nearest-neighbor alignment of the polarization field (Fig. 4.3), akin to velocity alignment in the Viscek model of collective motion (Vicsek et al. 1995). The active mechanochemical coupling $w > 0$ represents the rate of alignment of cell polarization with gradients in the actomyosin concentration field. As a result, local cell motion is guided toward regions of high contractility.

The concentration of contractile actomyosin is described by a reaction-advection-diffusion equation,

$$\begin{aligned}\partial_t c + \nabla \cdot (c\mathbf{v}) \\ = D\nabla^2 c - \frac{1}{\tau_c}(c - c_0) + \alpha c_0 \frac{\nabla \cdot \mathbf{u}}{1 + |\nabla \cdot \mathbf{u}|/s_0},\end{aligned} \tag{4.11}$$

where D is a diffusion constant, τ_c is the timescale of relaxation to steady-state, and $\alpha > 0$ is the rate of accumulation of contractile actomyosin due to local tissue stretching (Banerjee et al. 2015). The positive constant s_0 sets the upper limit of strain magnitude above which the production rate of c saturates (Köpf and Pismen 2013). This mechanochemical feedback (Fig. 4.4) is consistent with experimental data for single cells (Robin et al. 2018) and cell monolayers (Serra-Picamal et al. 2012; Vincent et al. 2015), where a local extensile strain reinforces contractility via assembly of actomyosin (Levayer and Lecuit 2012). Turnover

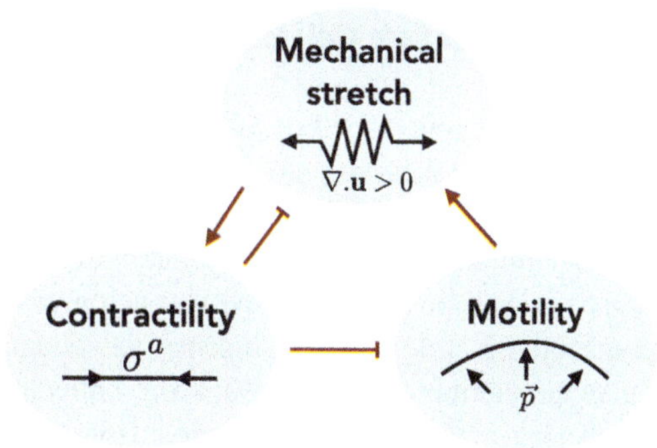

Fig. 4.4 Mechanochemical feedback mechanisms. Feedback between cell stretch, actomyosin contractility and polarized cell motility in the mechanochemical model for collective motion. Local stretch upregulates assembly of actomyosin, which generates contractile forces that exert compressive stress. Polarized motility, in turn, pulls and stretches the cells

of contractile elements at a rate τ_c^{-1} fluidizes the monolayer, inducing an effective viscosity of magnitude $\eta_{\rm eff} = (K - \sigma_0 + D\zeta/h)\tau_c$ (Banerjee et al. 2015). Aside from the negative feedback between mechanical strain and actomyosin assembly, positive feedback occurs between mechanical strain and advective fluxes into regions of high contractility. Advective transport can compete with diffusion to generate steady state patterns of contractility (Gross et al. 2017).

It is instructive to note that for small changes in c around c_0, Eq. (4.11) describes a dynamics of active contractile stress that is similar to a Maxwell constitutive model for intercellular stress proposed by Lee and Wolgemuth (2011). Here, in addition, we consider an elastic contribution to the active stress, described by the term α. The feedback between mechanical strain and contractility yields an effective elastic modulus $K_{\rm eff} \approx K + \alpha\tau_c(\sigma_0 + fw/2ah)$ (Banerjee et al. 2015), larger than the modulus K of the monolayer in the absence of contractility. This prediction is consistent with experimental measurements that cell monolayers treated with blebbistatin (myosin-II inhibitor) have a much reduced elastic modulus (Notbohm et al. 2016).

4.3 Forces and Motion Driving Collective Cell Behavior

The coupled system of Eqs. (4.8), (4.9), (4.10), and (4.11) describes the spatiotemporal dynamics of cell monolayers, subject to appropriate boundary and initial conditions for cellular displacement (**u**), cell polarity field (**p**) and actomyosin concentration (c). We now discuss the quantitative predictions of this model for collective mechanics and migration in various biological contexts. In particular we will focus on four scenarios where continuum model predictions have been tested and validated against experimental data: *Force transmission in epithelial monolayers* (Sect. 4.3.1), *Collective motility in expanding monolayers* (Sect. 4.3.2), *Cell migration under confinement* (Sect. 4.3.3), and *Epithelial movement during gap closure* (Sect. 4.3.4).

4.3.1 Force Transmission in Epithelial Monolayers

Epithelial cell monolayers adherent to soft elastic substrates provide a model system for mechanical force generation during tissue growth, migration and wound healing (Ladoux and Mège 2017; Wozniak and Chen 2009). In the experimental assays of interest (Du Roure et al. 2005; Trepat et al. 2009), the substrates are usually coated with extracellular matrix proteins (e.g. fibronectin, collagen) that allow cells to spread fully to a thin film and thereby establish contractile tension. To describe the experimentally observed traction force localization in fully spread adherent cell sheets (Du Roure et al. 2005; Mertz et al. 2012; Trepat et al. 2009), we consider the steady-state limit of Eqs. (4.8), (4.9), (4.10), and (4.11), which was originally studied in Edwards and Schwarz (2011), Banerjee and Marchetti (2011a, 2012), and Mertz et al. (2012). In this limit, $\mathbf{v} \equiv 0$, and the concentration of active contractile units is slaved to material strain, $c \approx c_0(1 + \alpha\tau_c\nabla\cdot\mathbf{u})$. This results in renormalization of the compressional modulus to linear order. Similarly from Eq. (4.10) it follows that $\mathbf{p} \approx -\left(\frac{w\alpha\tau_c c_0}{\kappa}\right)\mathbf{u}$.

To linear order, the force balance equation for the contracting cell layer, with internal stress $\boldsymbol{\sigma} = \boldsymbol{\sigma}^c + \sigma_0\mathbb{1}$, is given by,

$$h\nabla\cdot\boldsymbol{\sigma} = Y\mathbf{u}, \tag{4.12}$$

where, $Y = k + \frac{fw\alpha\tau_c c_0}{\kappa}$ is the effective substrate rigidity, resulting from the sum of substrate stiffness k, and the contribution from cell polarization. The intercellular stress, $\boldsymbol{\sigma}^c$, follows a constitutive relation identical to that of a linear elastic solid with a renormalized compressional modulus $K_{\rm eff}$. Equation (4.12) can be exactly solved for circularly shaped monolayers (Edwards and Schwarz 2011; Mertz et al. 2012), subject to the stress-free boundary condition: $\boldsymbol{\sigma}\cdot\hat{\mathbf{n}} = 0$, where $\hat{\mathbf{n}}$ is the unit normal to the boundary of the monolayer. This boundary condition needs to be appropriately modified if the colony edge is under tension due to peripheral actin structures (Ravasio et al. 2015a).

The resulting solution to Eq. (4.12) describes cell traction forces and displacements localized to the edge of the monolayer over a length scale $\ell_p = \sqrt{K_{\mathrm{eff}} h/Y}$, defined as the *stress penetration depth*. Furthermore, internal stresses in the monolayer, $\boldsymbol{\sigma}$, accumulate at the center of the monolayer, in agreement with experimental data (Fig. 4.5a–b) (Tambe et al. 2011; Trepat et al. 2009). The model can be solved numerically for monolayers of any geometry, and it predicts that traction stresses localize to regions of high curvature of the tissue boundary (Banerjee and Marchetti 2013). This was later confirmed experimentally by micropatterning adhesion geometries of non-uniform curvatures (Oakes et al. 2014). The model has been used to recapitulate a number of experimental observations (Banerjee and Marchetti 2011a, 2012, 2013; Mertz et al. 2012), including substrate rigidity dependence of traction stresses (Ghibaudo et al. 2008) and cell spread area (Chopra et al. 2011), traction stress dependence on cell geometry (Oakes et al. 2014), correlation between cell shape and mechanical stress anisotropy (Roca-Cusachs et al. 2008), as well as the optimal substrate rigidity for maximal cell polarization (Zemel et al. 2010).

A particularly interesting application of this model is in understanding the relationship between traction force magnitude and the geometric size of cohesive cell colonies adherent to soft matrices (Mertz et al. 2012). One can define the magnitude of the total traction force transmitted to substrate as $\mathscr{F} = \int |\mathbf{T} \cdot \mathbf{dA}|$, where the integral is taken over the entire spread area of the colony, A. The model predicts that for large cell colonies of linear size $R \gg \ell_p$, $\mathscr{F} = 2\pi h \sigma_0 R$. This linear scaling of force with colony size (Fig. 4.6) implies that actomyosin contractility, σ_0, induces an effective surface tension in solid tissues, which appear to wet the substrate underneath akin to fluid droplets. The effective surface tension was estimated from experiments on keratinocyte colonies to be 8×10^{-4} N/m (Mertz et al. 2012), which is of the same order of magnitude as the apparent surface tension estimated in adherent endothelial cells (Bischofs et al. 2009), Dictyostelium cells (Delanoë-Ayari et al. 2010), mm-scale migrating epithelial sheets (Trepat et al. 2009), and cellularised aggregates (Guevorkian et al. 2010). Recent work has shown that for highly motile and fluid cell colonies, traction forces localize to the colony interior rather than at the edge (Schaumann et al. 2018).

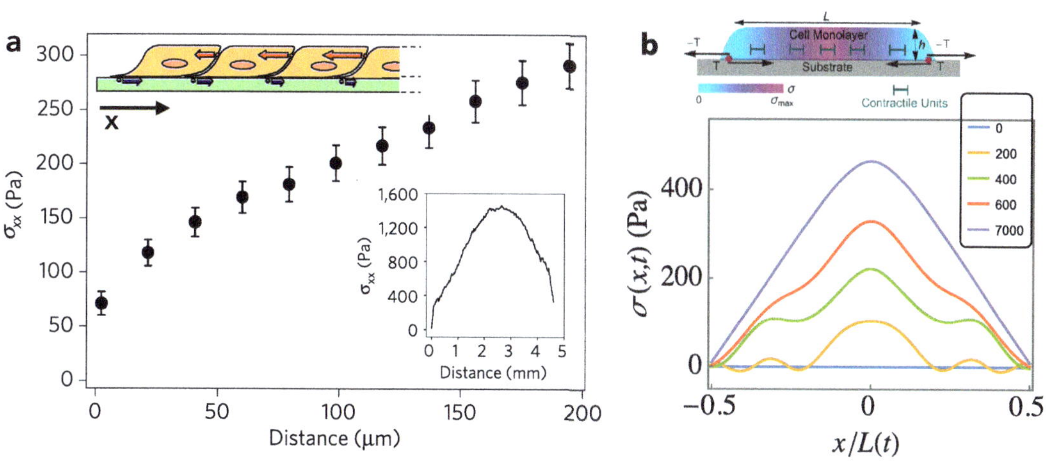

Fig. 4.5 Stress transmission in epithelial monolayers. (**a**) Internal stress, σ_{xx}, in an expanding MDCK monolayer obtained by integrating cellular traction force. (Adapted from Trepat et al. 2009). Buildup of σ_{xx} signifies that tension in the actin cytoskeleton and cell-cell junctions increases towards the centre of the monolayer. (**b**) Time evolution of the internal stress $\sigma(x, t)$ in the monolayer predicted by the continuum model of epithelium (Banerjee et al. 2015)

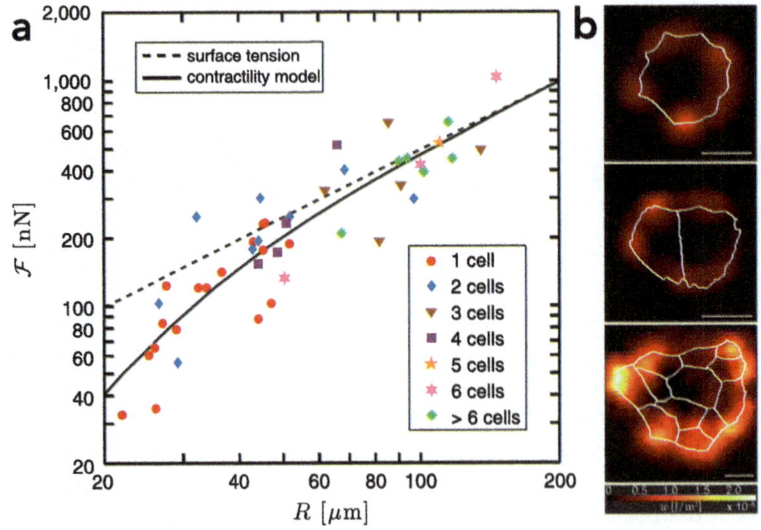

Fig. 4.6 **Active surface tension in cohesive epithelial colonies.** (**a**) Total force transmitted to the substrate by keratinocyte colonies, $\mathcal{F}$, as a function of the equivalent radius, R, of the colonies (Mertz et al. 2012). The dashed line represents the linear scaling expected for surface tension, $\mathcal{F} \propto R$. The solid line shows a fit of the data to the continuum model in Eq. (4.12). (**b**) Distribution of strain energy, w, for a representative single cell, pair of cells, and colony of 12 cells. Scale bar = 50 μm

4.3.2 Collective Motility in Expanding Monolayers

Migratory behaviors of epithelial cells are commonly studied experimentally using the wound healing assay. In the classical *scratch-assay* (Yarrow et al. 2004), a strip of cells is removed from the monolayer to observe collective migration of cells marching to fill the tissue gap. This experimental model system, however, is unsuited for controlled study of migration due to ill-defined borders and debris created by the physical wound. The last decade has seen significant improvement in the wound healing assay, where cells are grown to confluence within a removable barrier, which is then lifted to allow cell migration into free space (Poujade et al. 2007; Trepat et al. 2009). These studies, in combination with Traction Force Microscopy have shed light into the forces and motion driving collective cell migration. In particular, it has been observed that cell velocity fields at the leading edge of the epithelium exhibit complex swirling patterns (Petitjean et al. 2010) and often form multicellular *migration fingers* (Poujade et al. 2007). Measurement of mechanical stresses at cell-cell and cell-substrate interfaces have given rise to models of *tug-of-war* (Trepat et al. 2009), a consequence of mechanical force-balance, where local traction stresses in the monolayers are integrated into long-ranged gradients of intercellular tensions (Fig. 4.5a–b). Stress inference at cell-cell junctions have led to the suggestion of *plithotaxis* (Tambe et al. 2011), where cell migration is guided towards the direction of maximum normal stress and minimum shear stress.

A particularly interesting case is that of collective migration waves, observed in mm-sized monolayers expanding into free space (Serra-Picamal et al. 2012) (Fig. 4.7a). These mechanical waves, crucially dependent on myosin contractility and cell-cell adhesions, propagate at a slow speed (on the order of μm/hr) from the colony edge to the center and back (Fig. 4.7b). The waves are mediated by shape changes at the scale of single cells. Pulling forces from crawling cells at the leading edge of the colony stretch interior cells, which periodically recover their shape via a proposed model of cytoskeletal flu-

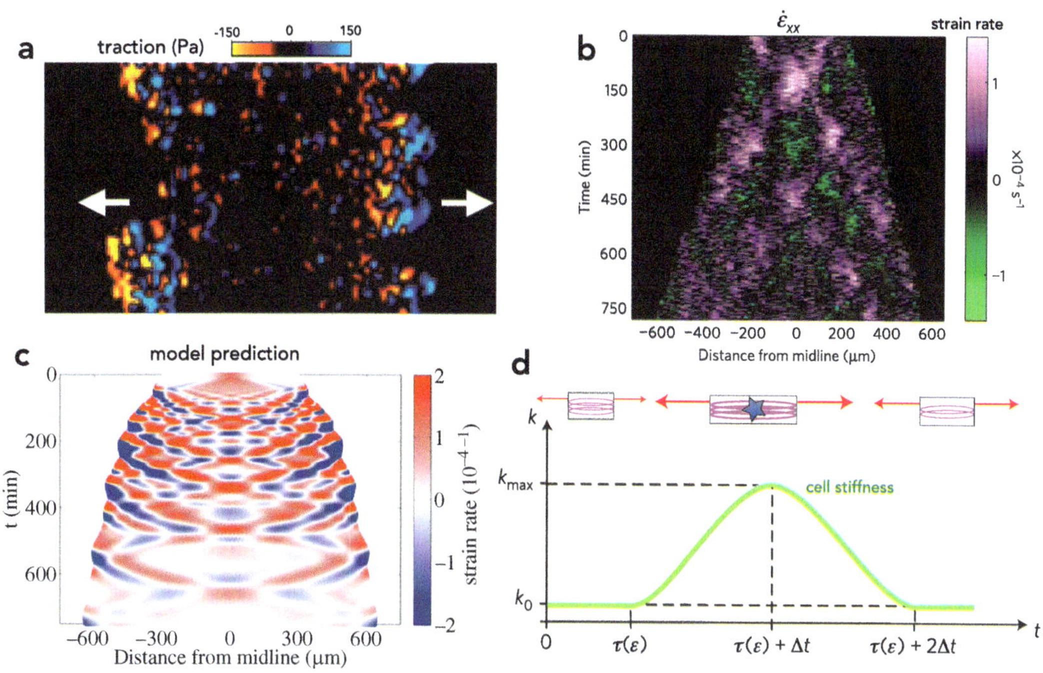

Fig. 4.7 Mechanical waves during epithelial expansion. (**a**) Traction stress map of an expanding MDCK monolayer. (Adapted from Serra-Picamal et al. 2012). (**b**) Kymograph of strain rate in expanding MDCK monolayers (Serra-Picamal et al. 2012), showing generation and propagation of X-shaped mechanical waves. (**c**) Propagating stress waves predicted by the continuum model, Eqs. (4.8), (4.9), (4.10), and (4.11) (Banerjee et al. 2015). (**d**) Schematic illustrating the mechanics of migration waves. (Adapted from Théry 2012 and Serra-Picamal et al. 2012). Cells at the colony center (purple) are initially stretched by pulling forces generated by leader cells. Stretched cells recover their equilibrium shape via cytoskeletal fluidization (blue star), which is then reinforced to trigger shape elongation again. These shape oscillations mediate periodic stiffening and fluidization of cells (green curve)

idization (Fig. 4.7d) (Théry 2012). Interestingly, this wave-like progression of cell movement naturally arises in the active elastic media models, Eqs. (4.8), (4.9), (4.10), and (4.11), due to a feedback between contractility and mechanical strain (Banerjee et al. 2015).

To understand the origin of wave propagation and estimate the wave frequency, it is useful to examine the mechanics of an expanding one-dimensional monolayer with a polarization field pointing outward from the colony center. We consider the linear fluctuations in the strain field $\delta\varepsilon$ and the concentration field δc, about the quiescent homogeneous state $u = 0$, $c = c_0$. Using Eqs. (4.8) and (4.11), one can eliminate δc to obtain the linearized dynamics of strain fluctuations:

$$\tau_c\zeta\,\partial_t^2\delta\varepsilon + \zeta\,\partial_t\delta\varepsilon = h(K_{\rm eff} + \eta_{\rm eff}\partial_t - \tau_c K D\partial_x^2)\partial_x^2\delta\varepsilon. \tag{4.13}$$

The above equation shows that the coupling of strain to concentration field yields an effective mass density (inertia), $\tau_c\zeta$, and viscoelasticity characterized by an effective elastic modulus, $K_{\rm eff}$, and an effective viscosity $\eta_{\rm eff}$, which leads to oscillations with a characteristic frequency $\omega = q\sqrt{h(K_{\rm eff} + \tau_c q^2 K D)/(\tau_c\zeta)}$, with q the wavevector. Full solutions of the nonlinear equations (Banerjee et al. 2015) yields X-shaped propagating stress waves akin to experimental data (Fig. 4.7c) (Serra-Picamal et al. 2012). These contraction waves are characterized by sustained oscillations in tissue rigidity – a slow period of stiffening followed by rapid fluidization (Fig. 4.7d). When the coupling of polarization to strain and contractility is turned on, complex spatiotemporal patterns emerge including traveling stress pulses and chaotic polarization waves (Banerjee et al. 2015; Köpf and Pismen 2013).

4.3.3 Cell Migration Under Confinement

In many biological contexts, including morphogenesis, tissue polarity establishment, and acini formation, cells often migrate collectively in confined environments. Experiments have shown the emergence of coherent rotation of cells in vivo (Fig. 4.8a), including cells in yolk syncytial layer of zebrafish embryos (D'Amico and Cooper 2001), and breast epithelial cells in 3D collagen gels (Tanner et al. 2012). These self-generated persistent motions are crucially dependent on cell-cell adhesions and myosin contractility, loss of which can drive malignant behavior. In recent years, collective motion in geometric confinement have been studied in a more controlled manner using adhesive micropatterns (Théry and Piel 2009), which allow confinement of cell cultures in geometric domains of any shape and size.

When plated in circular micropatterns, small sized epithelial monolayers often exhibit large scale correlated movements and spontaneous swirling motions, as shown in Fig. 4.8b (Deforet et al. 2014; Doxzen et al. 2013; Notbohm et al. 2016; Segerer et al. 2015). These collective rotations emerge once the cells have reached a critical density (2000 cells/mm^2) and occur in micropatterns of radii smaller than the cellular velocity correlation length ($\sim$200 μm) in unconfined situations (Doxzen et al. 2013). Furthermore these rotations require cell-cell adhesions for efficient transmission of motility cues by contact guidance (Doxzen et al. 2013), and radial velocity oscillations are observed with a time period linearly proportional to the micropattern radius (Deforet et al. 2014). Aside from collective rotational motion, emergence of active nematic states has also been observed in confined monolayers of elongated fibroblasts and MDCK cells (Duclos et al. 2014, 2017; Saw et al. 2017). In these cases, cells actively transfer alignment cues from the boundary to the bulk of the monolayer, resulting in domains of alignment and topological defect patterns.

Different cell-based computational models have been implemented to recapitulate collective rotational motion, including the cellular Potts model (Albert and Schwarz 2016; Doxzen et al. 2013; Kabla 2012), active particle models (Deforet et al. 2014), Vertex-based models (Schaumann et al. 2018), and Voronoi-type models (Li and Sun 2014), where persistent rotations emerge due to velocity alignment mechanisms of motile cells. In recent work (Notbohm et al. 2016), we described collective rotations using a continuum model similar to Eqs. (4.8), (4.9), (4.10), and (4.11) (Fig. 4.8c–d). This model quantitatively captures a key aspect of the experimental data, namely, that the cell velocity field alternated between inward and outward radial motion with a time period equal to that of the oscillations in the intercellular stress (Notbohm et al. 2016). This wave-like motion is predicted by the model to arise through the chemomechanical feedback between the mechanical strain, $\nabla \mathbf{u}$, and actomyosin contractility, c (Banerjee et al. 2015). In the limiting case where cell deformations, $\mathbf{u}$, are only coupled to polarity $\mathbf{p}$, no oscillatory behavior is observed. This prediction was confirmed by experiments, where inhibition of contractility by blebbistatin eliminated the multicellular oscillations. Furthermore, the polarization field, $\mathbf{p}$, is crucial to capture the misalignment between traction and velocity, observed experimentally. Overall, the coupling of cell motion to polarization and actomyosin contractility is required to capture the experimentally observed distribution of traction forces (Notbohm et al. 2016), which points inward at the periphery of the micropattern and oscillates between outward and inward within the bulk of the monolayer.

4.3.4 Epithelial Movement During Gap Closure

Collective cell movement during epithelial gap closure is essential for maintaining the tissue mechanical integrity and to protect the internal envi-

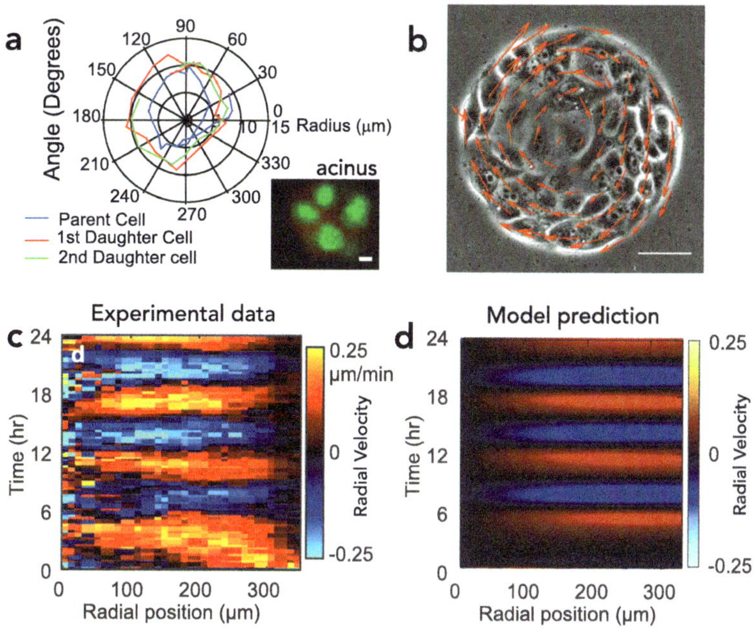

Fig. 4.8 Coherent cell motion in confined environment. (**a**) *Coherent angular motion* of cells during acinus morphogenesis. (Adapted from Tanner et al. 2012). Graph shows angular rotation of the parent and daughter cells obtained by nuclei tracking. Inset: Cross section of acinus with F-actin staining in green (Scale bar = 30 μm). (**b**) Collective rotation of MDCK cells seeded on circular fibronectin patterns. (Reproduced from Doxzen et al. 2013). The magnitude and the direction of local velocity fields are indicated by red arrows (Scale bar = 50 μm). (**c**) Kymograph of radial velocity fields of confluent cells in a micropattern (Notbohm et al. 2016), showing periodic oscillations. (**d**) Radial velocity kymograph, obtained by simulating Eqs. (4.8), (4.9), (4.10), and (4.11), reproducing collective cell oscillations

ronment from the outside by regenerating a physical barrier. Gaps can occur autonomously during development (Wood et al. 2002), or can be generated by cell apoptosis (Rosenblatt et al. 2001) or tissue injury. It is widely accepted that epithelial gap closure is driven by two distinct mechanisms for collective cell movement (Fig. 4.9a) (Begnaud et al. 2016; Jacinto et al. 2001). First, cells both proximal and distal to the gap can crawl by lamellipodial protrusions (Anon et al. 2012; Fenteany et al. 2000; Martin and Lewis 1992). Secondly, cells around the gap can assemble a multicellular actomyosin purse-string, which closes gaps via contractile forces (Fig. 4.9b) (Bement et al. 1993; Martin and Lewis 1992). The continuum framework described in this review can be appropriately adapted to study the relative contributions of crawling and contractile forces on epithelial gap closure.

Continuum models of tissue gap closure have considered both visco-elastic solid (Vedula et al. 2015) and fluid (Cochet-Escartin et al. 2014; Ravasio et al. 2015b) models of tissues. In either scenarios, force balance between cell-cell and cell-substrate interactions can be expressed as,

$$h\nabla \cdot \boldsymbol{\sigma} = \zeta \mathbf{v} - f\mathbf{p}, \qquad (4.14)$$

where f is the magnitude of the propulsion force acting on the cells due to lamellipodial protrusions, both proximal and distal to the gap, such that $\mathbf{p}$ points into free space. While previous continuum models have neglected the polarity term in the force balance, this is necessary for the misalignment of traction force and velocity observed for instance in closed contour wound healing assays (Brugués et al. 2014). To model the active pulling forces on the gap boundary, Eq. (4.14) is solved subject to the following boundary condition for the stress tensor on the moving gap boundary (Fig. 4.9c):

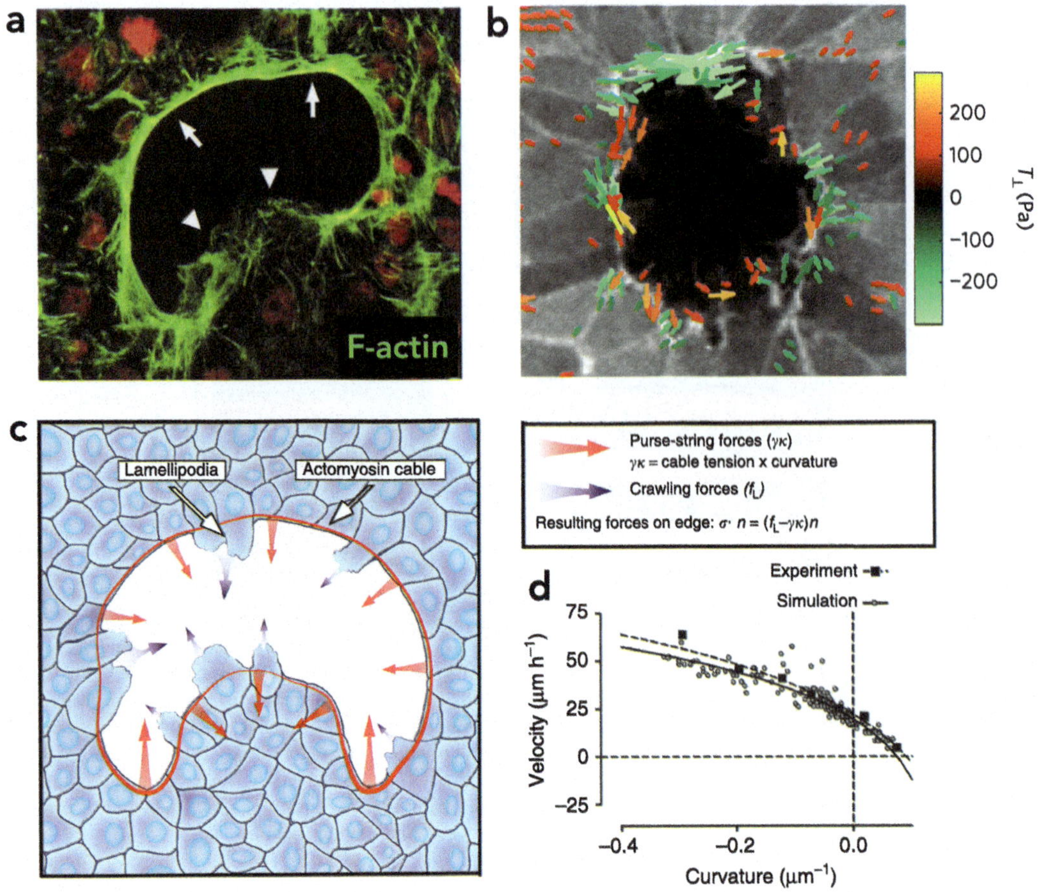

Fig. 4.9 Collective migration during epithelial gap closure. (**a**) Closure of in vitro wound in epithelial monolayers is mediated by a combination of purse-string based contraction of actomyosin cable (arrows) and lamellipodia based cell crawling (arrowheads). (Figure adapted from Jacinto et al. 2001). (**b**) Lamellipodial protrusions generate traction forces away from the wound (red arrows), whereas traction generated by purse-string based contraction point towards the wound (green arrows). (Traction stress map reproduced from Brugués et al. 2014). (**c**) Schematic of a continuum model for gap closure, showing the dependence of purse-string and crawling forces on the local gap geometry. (**d**) Migration velocity increases with increasing magnitude of local gap curvature. (Reproduced from Ravasio et al. 2015b)

$$\boldsymbol{\sigma} \cdot \hat{\mathbf{n}} = (f_L - \lambda\kappa)\hat{\mathbf{n}}, \tag{4.15}$$

where $\hat{\mathbf{n}}$ is the local normal vector on the gap boundary, directed away from the tissue, f_L is the force density due to lamellipodial protrusions, κ is the local gap boundary curvature (negative for circular gaps), and λ is the line tension due to actomyosin purse-string. The model has been used to capture the sensitivity of collective motion on the local gap geometry (Ravasio et al. 2015b) (Fig. 4.9d). For instance, crawling mediated migration ($\lambda = 0$) occurs at a speed independent of gap curvature, whereas purely purse-string driven motility ($f_L = 0$) increases with decreasing radius of curvature. This may explain why purse-string is not assembled for large wounds, as its driving force is inversely proportional to the gap diameter. A model of cable reinforcement, where tension $\lambda \propto \kappa$, has also been proposed to account for the experimentally observed increase in closure velocity and traction stress with time (Vedula et al. 2015). A more comprehensive model of gap closure dynamics with spatiotemporal variations in lamellipodia and purse-string forces (Fig. 4.9b) has recently been implemented using the vertex model (Ajeti et al. 2019; Staddon et al. 2018).

4.4 Comparisons Between Active Elastic and Fluid Models of Collective Cell Migration

Previous work has employed both elastic (Banerjee et al. 2015; Köpf and Pismen 2013) and fluid (Arciero et al. 2011; Blanch-Mercader and Casademunt 2017; Blanch-Mercader et al. 2017; Lee and Wolgemuth 2011; Pérez-González et al. 2018; Recho et al. 2016) models of epithelial cell sheet to describe the dynamics of epithelial expansion, as probed for instance in wound healing assays (Fig. 4.7a). Both models can account for traveling waves, as observed in experiments, provided the sheet rheology is coupled to internal dynamical degrees of freedom, such as contractile activity (elastic model Banerjee et al. 2015) or cell division or polarization (fluid model Recho et al. 2016 and Blanch-Mercader and Casademunt 2017). On the other hand, tissues can undergo fluidization/stiffening cycles, respond elastically or viscously on different times scales, and there is still no continuum model capable of capturing their rheology across all time scales.

In this section we compare the viscous and elastic continuum approaches for modeling cell monolayers, focusing on a one-dimensional ($1d$) model that allows for an analytical solution. The $1d$ calculation can also be directly compared to experiments such as those shown in Fig. 4.5a, where the monolayer properties are generally averaged over the direction transverse to that of mean motion. Denoting by x the direction of monolayer expansion, the in-plane force balance equation is simply given by

$$\zeta v_x = f p_x + h \partial_x \sigma, \tag{4.16}$$

where $\sigma = \sigma_{xx} = \sigma^c + \sigma^a$. In the absence of cell division and tissue growth, the volume of the monolayer remains approximately constant during expansion. This requires the product $L(t)h(t)$ to remain constant, where $L(t)$ and $h(t)$ are the monolayer width in the direction of expansion and the monolayer thickness at time t, respectively.

To illustrate the difference between the fluid and elastic models we examine below the accumulation of contractile stresses in an isotropic expanding monolayer, with vanishing net polarization, that was discussed for the fluid case in Blanch-Mercader et al. (2017). In contrast to Blanch-Mercader et al. (2017) we assume $\sigma^a =$ constant, to incorporate contractile cell activity. We neglect both nonlinear active stresses and spatiotemporal variations of the concentration c of contractile actomyosin. We additionally use a quasi-static approximation for the cell polarization that is assumed to relax on time scales much faster than those associated with cellular deformations and rearrangements. Finally, for simplicity we will neglect the spatial and temporal variation of the thickness h of the monolayer. We retain only linear terms so that the polarization field, p_x, satisfies the equation

$$L_p^2 \partial_x^2 p_x = p_x, \tag{4.17}$$

where we have introduced the length scale $L_p = \sqrt{\kappa/a}$ that describes spatial variation in polarization within the monolayer.

The viscous or elastic nature of the cell sheet will be specified by the chosen form of the constitutive equation for the intercellular stresses, σ^c. One important distinction, not apparent in the linear form of the equations considered here, is that the fluid motion is treated in an *Eulerian* frame, while the elastic medium model is implemented in a *Lagrangian* frame of reference. This difference will be important below when imposing boundary conditions.

The case of a fluid layer of growing width $2L(t)$ was discussed in Blanch-Mercader et al. (2017) (Fig. 4.10a). In this case intercellular stresses are purely viscous, with $\sigma^c = \eta \partial_x v_x$ and η the shear viscosity. Assuming that cells at the boundaries are outwardly polarized to drive expansion, i.e., $p_x(-L(t)) = -1$ and $p_x(+L(t)) = 1$, the static polarization profile is givenby

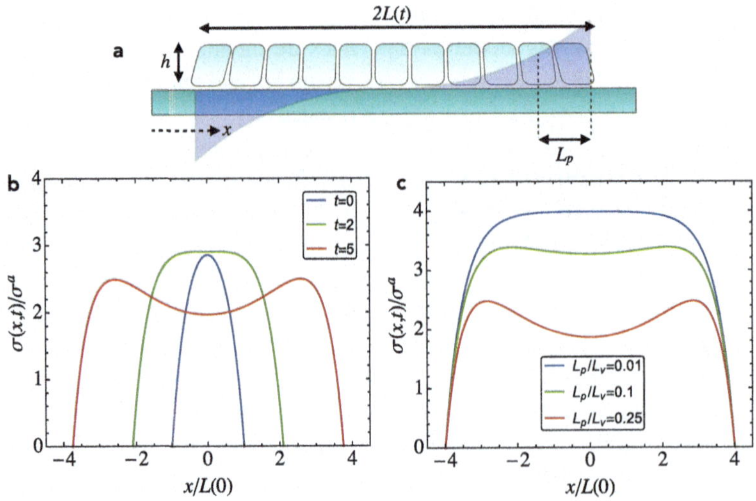

Fig. 4.10 Viscous fluid model of expanding monolayers. (**a**) Schematic of an expanding epithelial monolayer of height h and length $2L$, studied in Blanch-Mercader et al. (2017). The purple shaded curve represents the spatial profile of the polarization field, whose penetration depth is characterized by the length scale L_p. (**b**) Representative stress profiles of an expanding cell monolayer, predicted by the fluid model in Eq. (4.20), at different values of time with fixed $L_p/L_v = 0.25$. (**c**) Stress profiles for different values of L_p/L_v at fixed length $L = 4L(0)$. Other parameters: $L_p/L(0) = 0.5$, $f_0/\sigma^a = 2$

$$p_x(x) = \frac{\sinh(x/L_p)}{\sinh(L(t)/L_p)}. \tag{4.18}$$

The force balance equation, Eq. (4.16), can then be recast as an equation for the total stress in the fluid monolayer $\sigma_v(x) = \sigma(x)$,

$$\frac{1}{L_v^2}(\sigma_v - \sigma^a) = \frac{f_0}{L_p}\partial_x p_x + \partial_x^2 \sigma_v \tag{4.19}$$

where $L_v = \sqrt{h\eta/\zeta}$ is a viscous length scale, and $f_0 = fL_p/h$ is a characteristic stress scale. We solve Eq. (4.19) with stress-free boundary conditions at the monolayer edge, $\sigma_v(x = \pm L(t)) = 0$, where $L = L(t)$ is the growing monolayer length. The resultant stress is,

$$\sigma_v(x) = \sigma^a\left[1 - \frac{\cosh(x/L_v)}{\cosh(L/L_v)}\right] + \frac{f_0 L_v^2}{L_p^2 - L_v^2}\left[\frac{\cosh(x/L_p)}{\sinh(L/L_p)} - \frac{\cosh(L/L_p)\cosh(x/L_v)}{\sinh(L/L_p)\cosh(L/L_v)}\right]. \tag{4.20}$$

As shown in Blanch-Mercader et al. (2017) and in Fig. 4.10b–c, the shape of the stress profile depends on the length $L(t)$ as well as on the ratio $L_p/L_v > 0$. With increasing L (for fixed L_p/L_v) or increasing L_p/L_v (for fixed L), the initial stress maxima at the center of the layer disappears, and two stress peaks accumulate near the edge of the colony.

The length $L(t)$ of the expanding layer can be determined by equating the rate of change of $L(t)$ to the velocity at the leading edge, $\dot{L} = v_x(L)$. For $L(t) \gg L_p, L_v$ we find $v_x(L) \simeq f_0 L_v^2/\eta(L_p + L_v) - \sigma_a L_v/\eta$, resulting in a linear growth in time of the length of the monolayer,

$$L(t) = L_0 + \frac{L_p L_v^2}{h\eta(L_p + L_v)}(f - f_c^v)t, \tag{4.21}$$

provided the pulling force f exceeds a threshold value required to overcome the contractile force, $f_c^v = h\sigma^a(L_p^{-1} + L_v^{-1})$, and drive layer expansion. Note, however, that the assumption of indefinite growth in time in the absence of cell division is not realistic. Such a growth will be arrested by the requirement of volume conservation.

If the cell monolayer is modeled as an elastic continuum, then $\sigma^c = K\partial_x u_x$ where K is a compressional modulus and u_x the displacement field. The velocity must be identified with the rate of change of the displacement, $v_x = \partial_t u_x$. In this case the layer has a reference length $2L_0$ and an expanded length $2L(t) = 2L_0 + u(L_0, t) - u(-L_0, t)$. The polarization profile has the same functional form given in Eq. (4.18), but with $L(t)$ replaced by L_0. It is then evident that, in the absence of cell division and growth, the only steady state solution will have $v_x = 0$, corresponding to the fact that the elastic layer can be stretched by outward pulling cells, but not indefinitely expanded. The stress balance equation can again be written as a closed equation for the stress ($\sigma_{el}(x) = \sigma(x)$), $h\partial_x\sigma_{el} = -fp_x(x)$, with the solution (Fig. 4.5b)

$$\sigma_{el}(x) = f_0 \frac{\cosh(L_0/L_p) - \cosh(x/L_p)}{\sinh(L_0/L_p)}. \tag{4.22}$$

The stress profile of the elastically stretched tissue is controlled by the single length scale L_p and always shows a maximum at the midpoint of the layer. From this solution, one can immediately obtain the steady state displacement field, u_x, at the sample boundary, up to an undetermined constant. We eliminate this constant by assuming a symmetric deformation profile such that $u_x(0) = 0$. In the limit $L_0 \gg L_p$ we get $u_x(L_0) = -u_x(-L_0) = \frac{L_0}{K}\left(f_0 L_p/L_0 - \sigma^a\right)$. Of course in this case the monolayer stretches only provided the pulling forces due to polarization exceed the contractile forces. There is a critical pulling force, given by $f_c^e = h\sigma^a L_0/L_p^2$. Retaining again only leading terms in L_p/L_0, the total length of the expanded monolayer is then given by

$$L_\infty = L_0\left[1 + \frac{L_p^2}{L_0 h}(f - f_c^e)\right]. \tag{4.23}$$

Unlike the fluid monolayer, a purely elastic monolayer cannot sustain a state of steady growth. To obtain steady expansion in the case where the layer is modeled as an elastic medium it is necessary to include cell division. This can be accomplished in several ways: by allowing the reference layer length L_0 to grow with time; by describing cell division in terms of an extensile contribution to the active stress, such as $\sigma^{a,g} = -Rt$, where $R > 0$ describes the rate of growth; or by allowing the elastic constant K to vary in time. Each of these prescriptions will in general give different expansion rates for the monolayer. A full discussion of these cases is beyond the scope of the present review. In general, both viscous and elastic models have successfully reproduced the stress, velocity and deformation profiles measured in experiments. This suggests that these large scale quantities may not be terribly sensitive to the specific rheology of the monolayer. More work, however, remains to be done to fully understand the mechanisms that allow living tissues to maintain their cohesiveness, while exhibiting the fluidity necessary for motion and morphological changes, and to formulate a rheological model capable of capturing these unique properties.

4.5 Conclusion

Continuum models of multicellular mechanics have been widely successful in describing the physical forces, flow and deformation patterns that mediate collective cell migration during wound healing, tissue morphogenesis and development. These models are largely based on phenomenological approaches rooted in soft condensed matter physics, fluid dynamics and statistical mechanics (Marchetti et al. 2013). One of the key advantages of a continuum framework is that it is formulated in terms of a few coarse-grained *collective variables* such as density, velocity, strain and stress fields which are directly measurable in experiments. The resultant theory contains only a small number of macroscopic parameters, representing the effective mechanochemical couplings that arise from the combined effect of a number of signaling pathways at subcellular and cellular scales.

On the other hand, continuum models are generally written down phenomenologically, leaving open the key challenge of relating the continuum scale mechanical parameters to specific processes that control the active behavior of cells at μm and nm scales. In the absence of such a connection between subcellular and cellular or tissue scale, there are no constraints on the range of values spanned by the parameters of the continuum model. Many of the molecular pathways that mediate force generation and movement in cells are, however, intimately coupled and also sensitive to external perturbations and to the physical properties the cell's environment. It is then likely that molecular scale feedback processes may constrain the range of parameter values that are accessible at the cellular and tissue scales. As a result, all the complex dynamical phases predicted by generic continuum models may not be realizable in biological systems, as particular cells and tissues may likely operate in a narrow region of parameter space.

Another key limitation of the continuum modeling approach lies in the assumption of fixed materials properties of tissues, which is encoded in the choice of a particular constitutive law. As discussed elsewhere (Khalilgharibi et al. 2016), tissue rheology is highly complex, and the presence of multiple relaxation times demands a rheological model capable of capturing both active solid-like and fluid-like behavior in different regimes of stress response. In this review, we focus on active elastic models of tissue mechanics (Banerjee and Marchetti 2012; Banerjee et al. 2015; Edwards and Schwarz 2011; Köpf and Pismen 2013; Mertz et al. 2012; Notbohm et al. 2016) which have been successful in capturing many experimentally observed cell behaviors during collective migration. These include mechanical waves (Serra-Picamal et al. 2012), collective cell rotations (Deforet et al. 2014; Doxzen et al. 2013; Notbohm et al. 2016), traction force localization (Mertz et al. 2012; Trepat et al. 2009), and mechanosensitivity to extracellular matrix properties (Schwarz and Safran 2013). We also compare elasticity models against viscous fluid models of cell migration (Blanch-Mercader et al. 2017), showing that macroscopic quantities and observables may not sensitive to the specific choice of tissue rheology. On the other hand, a number of mesoscopic models, such as the Vertex, Voronoi, Potts and particle-based models, have been shown to capture various aspects of tissue-scale mechanics, providing an alternate bottom-up approach that may allow us to connect molecular scale to tissue-scale properties. A systematic study of such models with an eye on developing the multi-scale mechanics of multicellular assemblies is currently lacking, and remains an open theoretical challenge at the interface of physics and biology.

Living cells are active entities, capable for instance of autonomous motion, spontaneous mechanical deformations, division and phenotypical changes. This behavior can often be modeled at the mesoscale through internal state variables unique to living systems. In this review we have introduced two such internal state variables: the concentration of intracellular molecular active force generators and the cell polarity vector that describes the direction in which individual cells tend to move. For simplicity we have only considered the concentration of contractile units in the actomyosin cytoskeleton, that may represent, for instance, phosporylated myosins. More generally, several dynamically coupled chemical components may be needed to capture the complexity of molecular processes in the cell cytoskeleton. Multiple filaments, motors, and binding proteins compete to regulate cell homeostasis, polarization, and active force generation (Suarez and Kovar 2016). As more fascinating regulatory properties of the cytoskeletal machinery are being discovered, future models must attempt to incorporate such self-regulatory mechanisms controlling active cell mechanics.

An open question is the molecular interpretation of the cell polarization. Different interpretations have been put forward in the literature, including identifying cell polarization with the direction of lamellipodial/filopodial protrusions or with the orientation of the cell long axis associated with the alignment of actin stress fibers, although the latter provides a nematic (head-tail symmetric) degree of freedom, rather than a polar one. Regardless of its subcellular origin, cell polarity serves to dictate the direction of local motion, and is distinct from the actual direction of cell motion in a tissue that is also controlled by the forces from neighboring cells. In other words, the dynamics of the polarity vector encodes the decision-making rules for cell motility that come from the sum of mechanical and biochemical cues that an individual cell experiences from its internal as well as external environments. Given the multitude of polarity cues gathered by a cell, it remains contentious whether a single polarity state variable can fruitfully describe multiple mechanisms of active cell motility.

Essential ingredients of the models described in this review are the feedbacks between cellular mechanics, polarized motility, and the regulatory biochemistry of actomyosin contractility. Mechanochemical coupling of cell motion, adhesion and contractility have been argued as the physical basis for tissue morphogenesis and development (Howard et al. 2011). These couplings also play an essential role in the transmission of spatial information in large cell monolayers, mediated by waves, pulses, and a *tug of war* between cell-cell and cell-substrate forces. Both negative and positive feedback loops are exploited by cells for robust movement and force generation. Positive feedback commonly occurs between mechanical strain and advective transport of cytoskeletal filaments and motors into regions of high contractility. These active forces compete with diffusion and elasticity to establish the spatial gradients of contractility responsible for spontaneous cell motion. On the other hand, negative feedback between mechanical strain and contractility can yield periodic cycles of tissue stiffening and fluidization, which can result in long-range propagation of mechanical waves in tissues. At present, however, these feedback mechanisms remain purely phenomenological constructs, with only qualitative support from experiments. Their direct quantification is an outstanding experimental challenge.

In the future, theorists and experimentalists will need to work together to identify and probe all the key mechanical and biochemical parameters in a single model system. Such collaborative efforts will lead the way to more quantitatively accurate models of collective cell behavior in physiology and development.

Acknowledgements SB acknowledges support from a Strategic Fellowship at the Institute for the Physics of Living Systems at UCL, UCL Global Engagement Fund, Royal Society Tata University Research Fellowship (URF\R1\180187), and Human Frontiers Science Program (HFSP RGY0073/2018). MCM was supported by the National Science Foundation at Syracuse University through award DMR-1609208 and at KITP under Grant PHY-1748958, and by the Simons Foundation through a Targeted Grant Award No. 342354. MCM thanks M. Czajkowski for useful discussions and the KITP for hospitality during completion of some of this work.

References

Ahmadi A, Marchetti MC, Liverpool TB (2006) Hydrodynamics of isotropic and liquid crystalline active polymer solutions. Phys Rev E 74(6):061913

Ajeti V, Tabatabai AP, Fleszar AJ, Staddon MF, Seara DS, Suarez C, Yousafzai MS, Bi D, Kovar DR, Banerjee S, Murrell MP (2019) Wound healing coordinates actin architectures to regulate mechanical work. Nat Phys 5:696

Albert PJ, Schwarz US (2016) Dynamics of cell ensembles on adhesive micropatterns: bridging the gap between single cell spreading and collective cell migration. PLoS Comput Biol 12(4):e1004863

Angelini TE, Hannezo E, Trepat X, Fredberg JJ, Weitz DA (2010) Cell migration driven by cooperative substrate deformation patterns. Phys Rev Lett 104(16):168104

Angelini TE, Hannezo E, Trepat X, Marquez M, Fredberg JJ, Weitz DA (2011) Glass-like dynamics of collective cell migration. Proc Nat Acad Sci 108(12):4714

Anon E, Serra-Picamal X, Hersen P, Gauthier NC, Sheetz MP, Trepat X, Ladoux B (2012) Cell crawling mediates collective cell migration to close undamaged epithelial gaps. Proc Nat Acad Sci 109(27):10891

Arciero J, Mi Q, Branca MF, Hackam DJ, Swigon D (2011) Continuum model of collective cell migration in wound healing and colony expansion. Biophys J 100:535

Banerjee S, Marchetti MC (2011a) Substrate rigidity deforms and polarizes active gels. Europhys Lett 96(2):28003

Banerjee S, Marchetti MC (2011b) Instabilities and oscillations in isotropic active gels. Soft Matter 7(2):463

Banerjee S, Marchetti MC (2012) Contractile stresses in cohesive cell layers on finite-thickness substrates. Phys Rev Lett 109(10):108101

Banerjee S, Marchetti MC (2013) Controlling cell-matrix traction forces by extracellular geometry. New J Phys 15(3):035015

Banerjee S, Liverpool TB, Marchetti MC (2011) Generic phases of cross-linked active gels: relaxation, oscillation and contractility. Europhys Lett 96(5):58004

Banerjee S, Utuje KJ, Marchetti MC (2015) Propagating stress waves during epithelial expansion. Phys Rev Lett 114(22):228101
Banerjee DS, Munjal A, Lecuit T, Rao M (2017) Actomyosin pulsation and flows in an active elastomer with turnover and network remodeling. Nat Commun 8(1):1121
Barton DL, Henkes S, Weijer CJ, Sknepnek R (2017) Active vertex model for cell-resolution description of epithelial tissue mechanics. PLoS Comput Biol 13(6):e1005569
Basan M, Risler T, Joanny JF, Sastre-Garau X, Prost J (2009) Homeostatic competition drives tumor growth and metastasis nucleation. HFSP J 3(4):265
Basan M, Elgeti J, Hannezo E, Rappel WJ, Levine H (2013) Alignment of cellular motility forces with tissue flow as a mechanism for efficient wound healing. Proc Nat Acad Sci 110(7):2452
Begnaud S, Chen T, Delacour D, Mège RM, Ladoux B (2016) Mechanics of epithelial tissues during gap closure. Curr Opin Cell Biol 42:52
Bement WM, Forscher P, Mooseker MS (1993) A novel cytoskeletal structure involved in purse string wound closure and cell polarity maintenance. J Cell Biol 121(3):565
Bi D, Lopez J, Schwarz J, Manning ML (2015) A density-independent rigidity transition in biological tissues. Nat Phys 11(12):1074
Bi D, Yang X, Marchetti MC, Manning ML (2016) Motility-driven glass and jamming transitions in biological tissues. Phys Rev X 6(2):021011
Bischofs IB, Schmidt SS, Schwarz US (2009) Effect of adhesion geometry and rigidity on cellular force distributions. Phys Rev Lett 103(4):048101
Blanch-Mercader C, Casademunt J (2017) Hydrodynamic instabilities, waves and turbulence in spreading epithelia. Soft Matter 13(38):6913
Blanch-Mercader C, Vincent R, Bazellières E, Serra-Picamal X, Trepat X, Casademunt J (2017) Effective viscosity and dynamics of spreading epithelia: a solvable model. Soft Matter 13(6):1235
Bois JS, Jülicher F, Grill SW (2011) Pattern formation in active fluids. Phys Rev Lett 106(2):028103
Bove A, Gradeci D, Fujita Y, Banerjee S, Charras G, Lowe AR (2017) Local cellular neighborhood controls proliferation in cell competition. Mol Biol Cell 28(23):3215
Brugués A, Anon E, Conte V, Veldhuis JH, Gupta M, Colombelli J, Muñoz JJ, Brodland GW, Ladoux B, Trepat X (2014) Forces driving epithelial wound healing. Nat Phys 10(9):683
Camley BA, Rappel WJ (2017) Physical models of collective cell motility: from cell to tissue. J Phys D Appl Phys 50(11):113002
Chopra A, Tabdanov E, Patel H, Janmey PA, Kresh JY (2011) Cardiac myocyte remodeling mediated by N-cadherin-dependent mechanosensing. Am J Physiol Heart Circ Physiol 300(4):H1252
Cochet-Escartin O, Ranft J, Silberzan P, Marcq P (2014) Border forces and friction control epithelial closure dynamics. Biophys J 106(1):65
D'Amico LA, Cooper MS (2001) Morphogenetic domains in the yolk syncytial layer of axiating zebrafish embryos. Dev Dyn 222(4):611
Deforet M, Hakim V, Yevick HG, Duclos G, Silberzan P (2014) Emergence of collective modes and tri-dimensional structures from epithelial confinement. Nat Commun 5:3747
Delanoë-Ayari H, Rieu J, Sano M (2010) 4D traction force microscopy reveals asymmetric cortical forces in migrating Dictyostelium cells. Phys Rev Lett 105(24):248103
Discher DE, Janmey P, Wang Yl (2005) Tissue cells feel and respond to the stiffness of their substrate. Science 310(5751):1139
Doxzen K, Vedula SRK, Leong MC, Hirata H, Gov NS, Kabla AJ, Ladoux B, Lim CT (2013) Guidance of collective cell migration by substrate geometry. Integr Biol 5(8):1026
Duclos G, Garcia S, Yevick H, Silberzan P (2014) Perfect nematic order in confined monolayers of spindle-shaped cells. Soft Matter 10(14):2346
Duclos G, Erlenkämper C, Joanny JF, Silberzan P (2017) Topological defects in confined populations of spindle-shaped cells. Nat Phys 13(1):58
Du Roure O, Saez A, Buguin A, Austin RH, Chavrier P, Siberzan P, Ladoux B (2005) Force mapping in epithelial cell migration. Proc Nat Acad Sci 102(7):2390
Edwards CM, Schwarz US (2011) Force localization in contracting cell layers. Phys Rev Lett 107(12):128101
Farhadifar R, Röper JC, Aigouy B, Eaton S, Jülicher F (2007) The influence of cell mechanics, cell-cell interactions, and proliferation on epithelial packing. Curr Biol 17(24):2095
Farooqui R, Fenteany G (2005) Multiple rows of cells behind an epithelial wound edge extend cryptic lamellipodia to collectively drive cell-sheet movement. J Cell Sci 118(1):51
Fenteany G, Janmey PA, Stossel TP (2000) Signaling pathways and cell mechanics involved in wound closure by epithelial cell sheets. Curr Biol 10(14):831
Fletcher AG, Osterfield M, Baker RE, Shvartsman SY (2014) Vertex models of epithelial morphogenesis. Biophys J 106(11):2291
Foty RA, Forgacs G, Pfleger CM, Steinberg MS (1994) Surface tensions of embryonic tissues predict their mutual envelopment behavior. Phys Rev Lett 72(14): 2298
Friedl P, Gilmour D (2009) Collective cell migration in morphogenesis, regeneration and cancer. Nat Rev Mol Cell Biol 10(7):445
Ghibaudo M, Saez A, Trichet L, Xayaphoummine A, Browaeys J, Silberzan P, Buguin A, Ladoux B (2008) Traction forces and rigidity sensing regulate cell functions. Soft Matter 4(9):1836
Gonzalez-Rodriguez D, Bonnemay L, Elgeti J, Dufour S, Cuvelier D, Brochard-Wyart F (2013) Detachment and fracture of cellular aggregates. Soft Matter 9(7):2282
Graner F, Glazier JA (1992) Simulation of biological cell sorting using a two-dimensional extended Potts model. Phys Rev Lett 69(13):2013

Gross P, Kumar KV, Grill SW (2017) How active mechanics and regulatory biochemistry combine to form patterns in development. Ann Rev Biophys 46:337
Guevorkian K, Colbert MJ, Durth M, Dufour S, Brochard-Wyart F (2010) Aspiration of biological viscoelastic drops. Phys Rev Lett 104(21):218101
Guillot C, Lecuit T (2013) Mechanics of epithelial tissue homeostasis and morphogenesis. Science 340(6137):1185
Harris AR, Peter L, Bellis J, Baum B, Kabla AJ, Charras GT (2012) Characterizing the mechanics of cultured cell monolayers. Proc Nat Acad Sci 109(41):16449
Heisenberg CP, Bellaïche Y (2013) Forces in tissue morphogenesis and patterning. Cell 153(5):948
Honda H, Eguchi G (1980) How much does the cell boundary contract in a monolayered cell sheet?. J Theor Biol 84(3):575
Howard J, Grill SW, Bois JS (2011) Turing's next steps: the mechanochemical basis of morphogenesis. Nat Rev Mol Cell Biol 12(6):392
Jacinto A, Martinez-Arias A, Martin P (2001) Mechanisms of epithelial fusion and repair. Nat Cell Biol 3(5):E117
Kabla AJ (2012) Collective cell migration: leadership, invasion and segregation. J R Soc Interface p. rsif20120448
Khalilgharibi N, Fouchard J, Recho P, Charras G, Kabla A (2016) The dynamic mechanical properties of cellularised aggregates. Curr Opin Cell Biol 42:113
Khalilgharibi N, Fouchard J, Asadipour N, Yonis A, Harris A, Mosaffa P, Fujita Y, Kabla A, Baum B, Munoz JJ et al (2019) Stress relaxation in epithelial monolayers is controlled by the actomyosin cortex. Nat Phys 15:839
Köpf MH, Pismen LM (2013) A continuum model of epithelial spreading. Soft Matter 9(14):3727
Ladoux B, Mège RM (2017) Mechanobiology of collective cell behaviours. Nat Rev Mol Cell Biol 18(12):743
Latorre E, Kale S, Casares L, Gómez-González M, Uroz M, Valon L, Nair RV, Garreta E, Montserrat N, del Campo A, Ladoux B, Arroyo M, Trepat X (2018) Active superelasticity in three-dimensional epithelia of controlled shape. Nature 563(7730):203
Lecuit T, Lenne PF, Munro E (2011) Force generation, transmission, and integration during cell and tissue morphogenesis. Annu Rev Cell Dev Biol 27:157
Lee P, Wolgemuth CW (2011) Crawling cells can close wounds without purse strings or signaling. PLoS Comput Biol 7(3):e1002007
Legant WR, Choi CK, Miller JS, Shao L, Gao L, Betzig E, Chen CS (2013) Multidimensional traction force microscopy reveals out-of-plane rotational moments about focal adhesions. Proc Nat Acad Sci 110(3):881
Levayer R, Lecuit T (2012) Biomechanical regulation of contractility: spatial control and dynamics. Trends Cell Biol 22(2):61
Li B, Sun SX (2014) Coherent motions in confluent cell monolayer sheets. Biophys J 107(7):1532
Linsmeier I, Banerjee S, Oakes PW, Jung W, Kim T, Murrell MP (2016) Disordered actomyosin networks are sufficient to produce cooperative and telescopic contractility. Nat Commun 7:12615
Lomakin AJ, Lee KC, Han SJ, Bui DA, Davidson M, Mogilner A, Danuser G (2015) Competition for actin between two distinct F-actin networks defines a bistable switch for cell polarization. Nat Cell Biol 17(11):1435
Marchetti MC, Joanny JF, Ramaswamy S, Liverpool TB, Prost J, Rao M, Simha RA (2013) Hydrodynamics of soft active matter. Rev Mod Phys 85(3):1143
Martin P, Lewis J (1992) Actin cables and epidermal movement in embryonic wound healing. Nature 360(6400):179
Maruthamuthu V, Sabass B, Schwarz US, Gardel ML (2011) Cell-ECM traction force modulates endogenous tension at cell-cell contacts. Proc Nat Acad Sci 108(12):4708
Mertz AF, Banerjee S, Che Y, German GK, Xu Y, Hyland C, Marchetti MC, Horsley V, Dufresne ER (2012) Scaling of traction forces with the size of cohesive cell colonies. Phys Rev Lett 108(19):198101
Mertz AF, Che Y, Banerjee S, Goldstein JM, Rosowski KA, Revilla SF, Niessen CM, Marchetti MC, Dufresne ER, Horsley V (2013) Cadherin-based intercellular adhesions organize epithelial cell-matrix traction forces. Proc Nat Acad Sci 110(3):842
Murray J, Oster G (1984) Cell traction models for generating pattern and form in morphogenesis. J Math Biol 19(3):265
Murrell M, Oakes PW, Lenz M, Gardel ML (2015) Forcing cells into shape: the mechanics of actomyosin contractility. Nat Rev Mol Cell Biol 16(8):486
Noll N, Mani M, Heemskerk I, Streichan SJ, Shraiman BI (2017) Active tension network model suggests an exotic mechanical state realized in epithelial tissues. Nat Phys 13(12):1221
Notbohm J, Banerjee S, Utuje KJ, Gweon B, Jang H, Park Y, Shin J, Butler JP, Fredberg JJ, Marchetti MC (2016) Cellular contraction and polarization drive collective cellular motion. Biophys J 110(12):2729
Oakes PW, Banerjee S, Marchetti MC, Gardel ML (2014) Geometry regulates traction stresses in adherent cells. Biophys J 107(4):825
Pérez-González C, Alert R, Blanch-Mercader C, Gómez-González M, Kolodziej T, Bazellieres E, Casademunt J, Trepat X (2018) Active wetting of epithelial tissues. Nat Phys 15(1): 79
Petitjean L, Reffay M, Grasland-Mongrain E, Poujade M, Ladoux B, Buguin A, Silberzan P (2010) Velocity fields in a collectively migrating epithelium. Biophys J 98(9):1790
Phillips H, Steinberg M (1978) Embryonic tissues as elasticoviscous liquids. I. Rapid and slow shape changes in centrifuged cell aggregates. J Cell Sci 30(1):1
Poujade M, Grasland-Mongrain E, Hertzog A, Jouanneau J, Chavrier P, Ladoux B, Buguin A, Silberzan P (2007) Collective migration of an epithelial monolayer in response to a model wound. Proc Nat Acad Sci 104(41):15988

Prost J (1995) The physics of liquid crystals, vol 83. Oxford university press, Oxford
Prost J, Jülicher F, Joanny JF (2015) Active gel physics. Nat Phys 11(2):111
Ranft J, Basan M, Elgeti J, Joanny JF, Prost J, Jülicher F (2010) Fluidization of tissues by cell division and apoptosis. Proc Nat Acad Sci 107(49):20863
Ravasio A, Le AP, Saw TB, Tarle V, Ong HT, Bertocchi C, Mège RM, Lim CT, Gov NS, Ladoux B (2015a) Regulation of epithelial cell organization by tuning cell-substrate adhesion. Integr Biol 7(10):1228
Ravasio A, Cheddadi I, Chen T, Pereira T, Ong HT, Bertocchi C, Brugues A, Jacinto A, Kabla AJ, Toyama Y, et al (2015b) Gap geometry dictates epithelial closure efficiency. Nat Commun 6:7683
Recho P, Ranft J, Marcq P (2016) Soft Matter 12:2381
Robin FB, Michaux JB, McFadden WM, Munro EM (2018) J Cell Biol, 217(12):4230
Roca-Cusachs P, Alcaraz J, Sunyer R, Samitier J, Farré R, Navajas D (2008) Micropatterning of single endothelial cell shape reveals a tight coupling between nuclear volume in G1 and proliferation. Biophys J 94(12):4984
Roca-Cusachs P, Conte V, Trepat X (2017) Quantifying forces in cell biology. Nat Cell Biol 19(7):742
Rosenblatt J, Raff MC, Cramer LP (2001) An epithelial cell destined for apoptosis signals its neighbors to extrude it by an actin-and myosin-dependent mechanism. Curr Biol 11(23):1847
Saw TB, Doostmohammadi A, Nier V, Kocgozlu L, Thampi S, Toyama Y, Marcq P, Lim CT, Yeomans JM, Ladoux B (2017) Topological defects in epithelia govern cell death and extrusion. Nature 544(7649):212
Schaumann EN, Staddon MF, Gardel ML, Banerjee S (2018) Force localization modes in dynamic epithelial colonies. Mol Biol Cell 29(23):2835
Schwarz US, Safran SA (2013) Physics of adherent cells. Rev Mod Phys 85(3):1327
Segerer FJ, Thüroff F, Alberola AP, Frey E, Rädler JO (2015) Emergence and persistence of collective cell migration on small circular micropatterns. Phys Rev Lett 114(22):228102
Serra-Picamal X, Conte V, Vincent R, Anon E, Tambe DT, Bazellieres E, Butler JP, Fredberg JJ, Trepat X (2012) Mechanical waves during tissue expansion. Nat Phys 8(8):628
Shraiman BI (2005) Mechanical feedback as a possible regulator of tissue growth. Proc Natl Acad Sci 102(9):3318–3323
Staddon MF, Bi D, Tabatabai AP, Ajeti V, Murrell MP, Banerjee S (2018) Cooperation of dual modes of cell motility promotes epithelial stress relaxation to accelerate wound healing. PLoS Comput Biol 14(10):e1006502
Style RW, Boltyanskiy R, German GK, Hyland C, MacMinn CW, Mertz AF, Wilen LA, Xu Y, Dufresne ER (2014) Traction force microscopy in physics and biology. Soft Matter 10(23):4047
Suarez C, Kovar DR (2016) Internetwork competition for monomers governs actin cytoskeleton organization. Nat Rev Mol Cell Biol 17(12):799
Tambe DT, Hardin CC, Angelini TE, Rajendran K, Park CY, Serra-Picamal X, Zhou EH, Zaman MH, Butler JP, Weitz DA et al (2011) Collective cell guidance by cooperative intercellular forces. Nat Mater 10(6):469
Tanner K, Mori H, Mroue R, Bruni-Cardoso A, Bissell MJ (2012) Coherent angular motion in the establishment of multicellular architecture of glandular tissues. Proc Nat Acad Sci 109(6):1973
Théry M (2012) Cell mechanics: Wave of migration. Nat Phys 8(8):583
Théry M, Piel M (2009) Adhesive micropatterns for cells: a microcontact printing protocol. Cold Spring Harb Protoc 2009(7):pdb
Trepat X, Wasserman MR, Angelini TE, Millet E, Weitz DA, Butler JP, Fredberg JJ (2009) Physical forces during collective cell migration. Nat Phys 5(6):426
Vedula SRK, Peyret G, Cheddadi I, Chen T, Brugués A, Hirata H, Lopez-Menendez H, Toyama Y, De Almeida LN, Trepat X, et al (2015) Mechanics of epithelial closure over non-adherent environments. Nat Commun 6:6111
Vicsek T, Czirók A, Ben-Jacob E, Cohen I, Shochet O (1995) Novel type of phase transition in a system of self-driven particles. Phys Rev Lett 75(6):1226
Vincent R, Bazellières E, Pérez-González C, Uroz M, Serra-Picamal X, Trepat X (2015) Active tensile modulus of an epithelial monolayer. Phys Rev Lett 115(24):248103
Walcott S, Sun SX (2010) Cytoskeletal cross-linking and bundling in motor-independent contraction. Proc Nat Acad Sci 107(17):7757
Wayne Brodland G, Wiebe CJ (2004) Mechanical effects of cell anisotropy on epithelia. Comput Methods Biomech Biomed Eng 7(2):91
Wood W, Jacinto A, Grose R, Woolner S, Gale J, Wilson C, Martin P (2002) Wound healing recapitulates morphogenesis in Drosophila embryos. Nat Cell Biol 4(11):907
Wozniak MA, Chen CS (2009) Mechanotransduction in development: a growing role for contractility. Nat Rev Mol Cell Biol 10(1):34
Yabunaka S, Marcq P (2017) Cell growth, division, and death in cohesive tissues: A thermodynamic approach. Phys Rev E 96(2):022406
Yarrow JC, Perlman ZE, Westwood NJ, Mitchison TJ (2004) A high-throughput cell migration assay using scratch wound healing, a comparison of image-based readout methods. BMC Biotechnol 4(1):21
Zemel A, Rehfeldt F, Brown A, Discher D, Safran S (2010) Optimal matrix rigidity for stress-fibre polarization in stem cells. Nat Phys 6(6):468
Ziebert F, Swaminathan S, Aranson IS (2011) Model for self-polarization and motility of keratocyte fragments. J R Soc Interface p. rsif20110433

Statistical Features of Collective Cell Migration

5

Caterina A. M. La Porta and Stefano Zapperi

Abstract

We discuss recent advances in interpreting the collective dynamics of cellular assemblies using ideas and tools coming from the statistical physics of materials. Experimental observations suggest analogies between the collective motion of cell monolayers and the jamming of soft materials. Granular media, emulsions and other soft materials display transitions between fluid-like and solid-like behavior as control parameters, such as temperature, density and stress, are changed. A similar jamming transition has been observed in the relaxation of epithelial cell monolayers. In this case, the associated unjamming transition, in which cells migrate collectively, is linked to a variety of biochemical and biophysical factors. In this framework, recent works show that wound healing induce monolayer fluidization with collective migration fronts moving in an avalanche-like behavior reminiscent of intermittent front propagation in materials such as domain walls in magnets, cracks in disordered media or flux lines in superconductors. Finally, we review the ability of discrete models of cell migration, from interacting active particles to vertex and Voronoi models, to simulate the statistical properties observed experimentally.

Keywords

Jamming · Bursts · Particle image velocimetry · Active particle models · Wound healing

5.1 Introduction

Collective cell migration is a fascinating topic of great biological relevance (Friedl and Gilmour 2009). Cells in tissues often do not move independently, but interact closely and move together. This phenomenon is relevant for cancer metastasis (La Porta and Zapperi 2017), where group of cells have been observed to collectively invade neighboring tissues (Gov 2014; Khalil and Friedl 2010; Rørth 2009). The biophysical aspects of collective cell migration can be discussed from the point of view of active matter (Ramaswamy 2010): Cell assemblies can be seen as peculiar type of out of equilibrium material that is able to convert internal biochemical energy into mechan-

C. A. M. La Porta (✉)
Department of Environmental Science and Policy, Center for Complexity and Biosystems, University of Milan, Milan, Italy
e-mail: caterina.laporta@unimi.it

S. Zapperi (✉)
Department of Physics, Center for Complexity and Biosystems, University of Milan, Milan, Italy
e-mail: stefano.zapperi@unimi.it

C. A. M. La Porta, S. Zapperi (eds.), *Cell Migrations: Causes and Functions*, Advances in Experimental Medicine and Biology 1146,
https://doi.org/10.1007/978-3-030-17593-1_5

ical and kinetic energy. As for ordinary matter, active matter can display transitions into different states with characteristic macroscopic properties in terms of flow or spatial correlations. The main aspect that we wish to explore in this chapter is related to the fluctuations associated to these states.

Individual cells move in a very erratic manner, performing a persistent random walk with statistical properties accurately described by simple stochastic differential equations. The problem becomes more intriguing when the cell concentration is increased and cells respond due to their mutual interactions. When cells are crowded they slow down, up to a point where their motion becomes confined as in a glass (Angelini et al. 2010; Park et al. 2015). This can be quantified by cell mean-square displacements that grow in diffusive or ballistic fashion for isolated cells and becomes bounded in crowded conditions (Malinverno et al. 2017). Cellular self-propulsion can counteract the caging effects due to crowding leading to a collectively flowing state. The most widespread interpretation of this phenomenon is in terms of the jamming/unjamming transition, widely observed in soft matter systems such as foams, colloids or granular media (Liu et al. 2010). The main difference for cells lies in the presence of internal active forces driving the transition and thus creating a completely new playground.

Self-propulsion forces become important in particular conditions, for instance when cells are faced with an empty space to invade, as in the case of wound healing (Chepizhko et al. 2018). The intermittent dynamics of an invading cell front is reminiscent to other fronts studied in condensed matter systems, such as crack lines or magnetic domain walls. The analogy is not only qualitative, since the distribution of bursts in cell front invasion follow the same statistical distributions as in disordered elastic systems in condensed matter (Chepizhko et al. 2016). Here, we discuss analogies and differences between cell migration and the transitions to flow in ordinary matter, focusing on few relevant experiments and on computational models based on interacting cells, represented as active particle (Berthier 2014; Fily et al. 2014; Flenner et al. 2016; Henkes et al. 2011; Liao and Xu 2018; Mandal et al. 2016; Poujade et al. 2007; Sepúlveda et al. 2013; Szabó et al. 2006; Szamel 2016; Vedula et al. 2013) or polygons in vertex models (Bi et al. 2015, 2016).

5.2 Fluctuations in the Migration of Individual Cells

Before discussing the fluctuations in the migration of collective assemblies of cells, it is useful to briefly recall here the stochastic behavior observed in individual cells as they migrate. It has been widely reported that cell trajectories in vitro display random fluctuations similar to those observed in Brownian particles (Codling et al. 2008; Potdar et al. 2010). Cells, however, are not just particles driven by the fluctuations in the fluid but involve internal active forces. As indeed shown by careful quantitative analysis, cells do not perform a simple random walk (Dieterich et al. 2008; Li et al. 2008; Potdar et al. 2010; Stokes et al. 1991; Wu et al. 2014) but a persistent random walk (PRW), characterized by long periods of persistent directional motion in one direction separated by re-orientation event, as illustrated in Fig. 5.1a (Li et al. 2008). This process is well described by a persistent random walk, following a simple Langevin equation (Stokes et al. 1991; Wu et al. 2014)

$$\frac{d\mathbf{v}}{dt} = -\frac{\mathbf{v}}{\tau} + \sqrt{\frac{D}{\tau}}\eta(t), \tag{5.1}$$

where $\mathbf{v}$ is the cell velocity, τ is the persistence time, $\eta(t)$ is an uncorrelated Gaussian noise with zero mean and unit variance, and the noise strength is tuned by D. This linear stochastic model can be easily solved and yields a mean-square displacement

$$\langle(\mathbf{r}(t+t_0) - \mathbf{r}(t_0))^2\rangle = 2D\tau\left(\exp(-t/\tau) + \frac{t}{\tau} - 1\right). \tag{5.2}$$

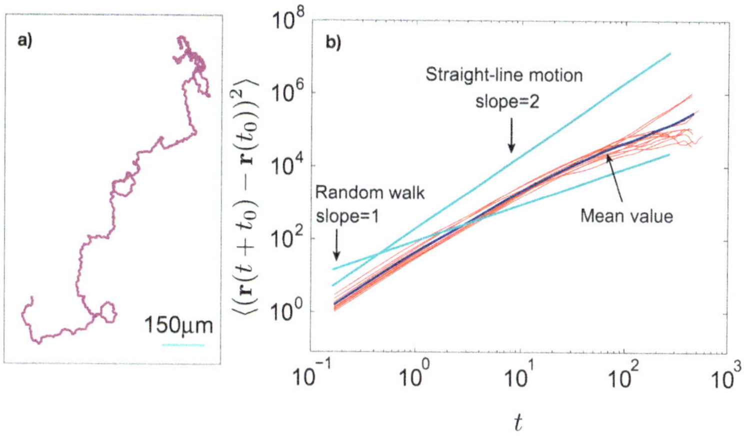

Fig. 5.1 (**a**) A trajectory of a Dictyostelium cell lasting for 10 h. (**b**) The mean square displacements recorded of Dictyostelium cells follow the prediction of the persistent random walk model. (Images adapted from Li et al. 2008) (Figs. 5.2a and 5.3b, CC licence)

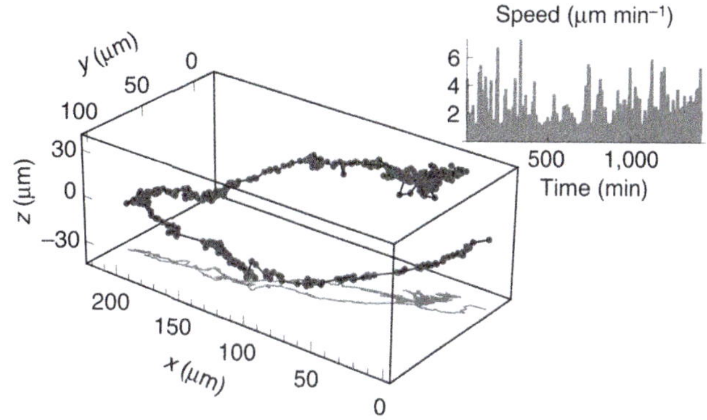

Fig. 5.2 A trajectory of a breast carcinoma cell migrating into a three dimensional collagen matrix . The inset shows the intermittent fluctuations of the cell velocity. (Image from Metzner et al. (2015) CC licence)

Equation 5.2 interpolates from an exponential increase at short times to a linear diffusive behavior at large times which agrees with experimental data for two dimensional motion as shown in Fig. 5.1b (Wu et al. 2014).

An alternative model to explain the deviation from pure Brownian motion is provided by anomalous diffusion, where the mean square displacement scales as $t^{2\alpha}$, with $\alpha > 1/2$ (Dieterich et al. 2008). Indeed, recent experiments tracked individual cells moving through a three dimensional collagen matrix (see Fig. 5.2) and showed clear deviations from the simple PRW model (Metzner et al. 2015; Wu et al. 2014). In particular, the distribution of velocities is not Gaussian as assumed in the PRW (Wu et al. 2014). The correct distribution can be obtained by modeling cell heterogeneity and substrate anisotropies (Wu et al. 2014) and introducing a superstatistical framework (Metzner et al. 2015) where the motion is modeled by a PRW with parameters (e.g. τ and D) that are themselves random functions. The superstatistical model is in excellent agreement with experimental results for cell motion in two and three dimensions (Metzner et al. 2015).

5.3 Avalanches and Fluctuations in Collective Cell Migration

Understanding collective cell migration, when cells move as a cohesive and coordinated group is important to shed light on key aspects of embryogenesis, wound repair and cancer metastasis (Friedl and Gilmour 2009). While cellular and multicellular dynamics and motility is controlled by a complex network of biochemical pathways (Ilina and Friedl 2009), it is becoming increasingly clear that a crucial role is also played by physical interactions among cells and between cells and their environment (Brugues et al. 2014; Haeger et al. 2014; Koch et al. 2012; Lange and

Fabry 2013; Tambe et al. 2011). In particular, experiments revealed that collective cell migration depends on the composition and stiffness of the extracellular matrix (ECM) on which the cells move (Brugues et al. 2014; Haeger et al. 2014; Koch et al. 2012; Lange and Fabry 2013; Tambe et al. 2011). The ECM of animal tissues is composed by a random hierarchical assembly of collegen fibrils and fibers whose mechanical properties have many advantages thanks to their characteristic non-linear strain stiffening, allowing for easy remodeling and high sensitivity at small deformations and higher rigidity against strong deformations (Sacks and Sun 2003). Collective cell migration can be studied in vitro by wound healing assays (Poujade et al. 2007; Sepúlveda et al. 2013; Szabó et al. 2006; Vedula et al. 2013), where confluent cell layers are scratched and the ensuing migration is observed in time lapse microscopy. When those studies are performed on substrates covered with collagen (Haga et al. 2005) and other gels (Ng et al. 2012) or micropatterned (Röttgermann et al. 2014; Saez et al. 2007), cell migration is found to crucially depend on the substrate structure and stiffness.

The statistical properties of collective cell migration do not only depend on the interaction between cells and substrate but also from the mutual interactions among cells. Experiments showed that cells are able to transfer mechanical stresses to their neighbors (Tambe et al. 2011), producing long-ranged stress waves in the monolayer (Banerjee et al. 2015; Serra-Picamal et al. 2012). This observation suggests an analogy with disordered elastic systems in materials, where the dynamics is ruled by the interplay of elastic interactions and the interaction with a quenched random field. In the case of collective cell migration, the elastic interactions are provided by intracellular adhesion, while the role of the random field is played by the substrate. Disordered elastic systems in materials, such as cracks lines (Maloy et al. 2006; Tallakstad et al. 2011), imbibition fronts (Clotet et al. 2014) or ferromagnetic domain walls (Durin and Zapperi 2000), all share common features. When the driving force (e.g. the external load for cracks, the fluid pressure for imbibition and the magnetic field in ferromagnets) overcomes a threshold value the system flows while at low forces it is pinned by the disorder. The depinning threshold is associated with a non-equilibrium critical point characterized by scaling laws for the statistical properties of the dynamics, as revealed by numerical simulations (Leschhorn et al. 1997; Rosso et al. 2009) and renormalization group theory (Chauve et al. 2001; Le Doussal and Wiese 2009; Leschhorn et al. 1997; Narayan and Fisher 1992). In particular close to the depinning threshold, the dynamic of the front is strongly fluctuating and intermittent, characterized by bursts of activity or avalanches. The statistics of these avalanche events follows a power law distribution with an exponent that is universal (i.e. it does not depend on the microscopic features of the system but only on the general symmetry of the interactions).

In a recent paper (Chepizhko et al. 2016), we have shown by a careful analysis of time-lapse imaging during wound healing that a migrating cell front shares many similarities with moving front close to the depinning transition. The analysis has been performed on a variety of cell lines (human cancer cells and epithelial cells, mouse endothelial cells) over different substrates (plastic, soluble and fibrillar collagen) and with varying experimental conditions (such as the tuning of intracellular adhesion by VE-cadherin knock down). An example of the evolution of the cell front in a monolayer of HeLa cell is reported in the bottom part of Fig. 5.3a. The fronts are rough and advance in bursts, as it is apparent by looking at activity map the top part of Fig. 5.3a, where the colored region corresponds to areas that move collectively, denoted as clusters of activity. Activity maps were obtained using an algorithm devised to study avalanches in planar crack propagation (Tallakstad et al. 2011) and imbibition (Clotet et al. 2014). As in the case of fracture or imbibition, the distribution of cluster areas S decays as a power law $P(S) \sim S^{-\tau}$ up to a cutoff length S^*, as illustrated in Fig. 5.3b for a variety of cell lines. It is interesting to remark that the value of the exponent $\tau \simeq 1.5$ is independent on the cell line (Chepizhko et al. 2016) and is

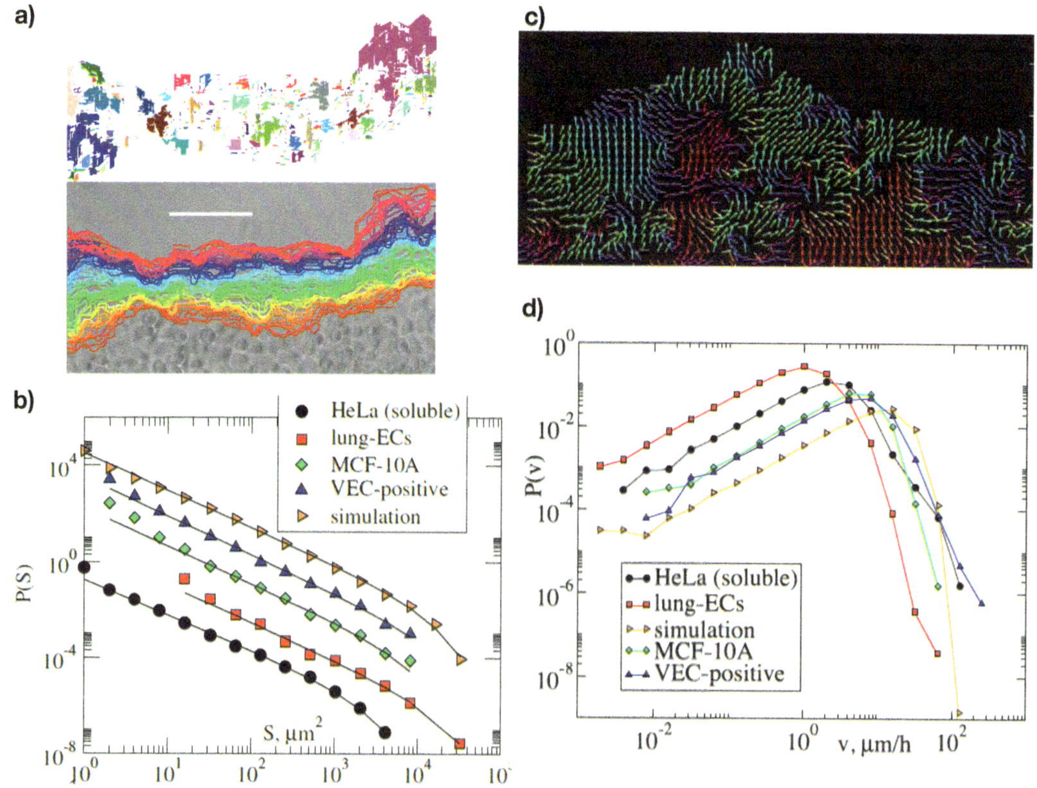

Fig. 5.3 Dynamic fluctuations in wound healing experiments. (**a**) An example of cell a cell front and the activity maps reconstructed from the time evolution of the front in Hela cells moving on a collagen substrate. Regions marked by the same color in the activity map move collectively. The scale bar is 100 μm. (**b**) Distributions of the areas of activity clusters display power law scaling with a cutoff. The distributions for different cell types have been shifted for clarity. The slope obtained fitting the distributions is very similar for all cell types. (**c**) Velocity map obtained from particle image velocimetry. The length of the arrows is proportional to the magnitude of the velocity. (**d**) Distributions of velocity magnitudes for different cell types. (Reprinted from Chepizhko et al. (2016) with permission)

similar to the one observed in fracture (Tallakstad et al. 2011).

In addition to the activity map, a useful technique to characterize the fluctuations in the dynamics of cell migration, both in confluent and in wound healing conditions, is provided by particle image velocimetry (PIV). PIV estimates local velocities by performing a digital image correlation analysis on the time-lapse sequence and allows to obtain a velocity map as the one reported in Fig. 5.3c for HeLa cells. The figure shows that cells move with significant fluctuations, also involving local motion that is opposite to the propagation direction of the front. The fluctuations can be captured by measuring velocity distributions as the ones reported in Fig. 5.3d. The distribution vary slightly for different cell lines, but the shape of the distribution is similar in all cases.

5.4 The Jamming/Unjamming Transition in Cell Assemblies

The flow behavior of a wide class of soft matter systems, from colloidal suspensions (Bonn et al. 2002) to emulsions (Mason et al. 1996), foams (Durian et al. 1991), gels (Segrè et al. 2001) and pastes (Cloitre et al. 2003) is ruled by the presence of kinematic constraints, leading to jamming, a concept describing the suppression of temporal relaxation and the corresponding ability to explore the space of configurations (Liu et al. 2010). These soft matter systems are typically composed randomly arranged particles, whose individual motion becomes constrained as the density increases. As a result of this, a jammed system responds like an elastic solid upon the application of low shear stresses. Under the action of externally applied shear stresses, however, these systems eventually yield and are able to flow like a viscous fluid. The yield stress depends also on the density, hindering the motion, and on the temperature that promotes flow. These observations can be summarized into a generic phase diagram for jamming systems that is reported in Fig. 5.4.

It has been argued that the collective dynamics of dense cellular assemblies, such as epithelial monolayers or cancer cell colonies, can bee described by the same framework employed for disordered soft matter. In particular, experiments show that cellular assemblies display slow glassy relaxation (Angelini et al. 2011) leading to a jammed state, characterized by limited cellular motility (Park et al. 2015). Depending on the experimental conditions, cells can collectively flow like a fluid, but mutual crowding typically leads to slowing down and dynamic arrest in a way that is similar to the behavior observed in soft matter across the jamming transition (Angelini et al. 2011; Park et al. 2015). As in disordered solids, cell jamming can occur across different routes, but the potential ways are clearly more diverse in living systems than in soft matter. For instance, cell jamming can be triggered by an increase of cell density as in conventional soft matter, or by other cell specific mechanism such the reduction of the active forces responsible for cell motility (Doxzen et al. 2013), increased intracellular adhesion (Garcia et al. 2015) or the expression level of some particular gene (Malinverno et al. 2017). For instance, Malinverno et al. (2017) showed that the over-expression of RAB5A, a master regulator of endocytosis, leads to rapid fluidization of a jammed epithelial layer. This unjamming is thought to arise due to the polarization of cell protrusion and increase in traction force. The role of cell-cell adhesion in affecting the properties of collective cell migration and the associated mechanical forces has been investigated extensively by knocking down more than twenty individual adhesion molecules (Bazellières et al. 2015). The results show that P-cadherin is related to the strength of the adhesion forces, while E-cadherin controls the rate at which force grow.

Understanding cell jamming is important, not only for the intriguing analogies with soft matter systems but also for its possible functional

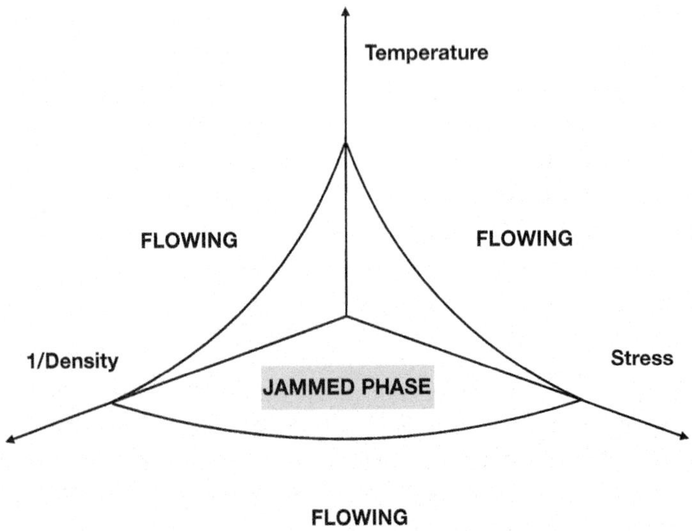

Fig. 5.4 A schematic jamming phase diagram in soft matter systems

biological role. Jamming could assist the development of tissue elasticity and the formation protective barriers in epithelial tissues, as well as suppressive mechanisms for the aberrant growth of cancer cells. While experiments clearly show that jamming is a relevant concept to describe the behavior of cellular assemblies, it is difficult to fine tune parameters to carefully study the transition. This, however, can be done resorting to theoretical and computational models, as we discuss in the next section.

5.5 Discrete Models for Collective Cell Migration

To understand the statistical properties of collective cell migration it is useful to resort to theoretical and computational models. Models are interesting because they not only allow to reproduce with minimal ingredients the main features of the experiments, but mostly because they allow to identify the basic biophysical mechanisms ruling the observed behavior. Furthermore, simulations allow to explore the role of various biophysical parameters in determining the migration properties of the assembly and sometimes to reconstruct a possible phase diagram. The theoretical and computational literature on the subject is rather vast and here we restrict our attention to two main classes of discrete models, based on interacting active particles or Voronoi lattices. We refer the reader to Chap. 4 for a detailed discussion of continuous models (Marchetti and Banerjee 2019).

5.5.1 Interacting Active Particles

In active particle models (Berthier 2014; Fily et al. 2014; Flenner et al. 2016; Henkes et al. 2011; Liao and Xu 2018; Mandal et al. 2016; Poujade et al. 2007; Sepúlveda et al. 2013; Szabó et al. 2006; Szamel 2016; Vedula et al. 2013), cells are modeled as a set of particles mutually interacting through an attractive force, due to intra-cellular adhesion, and hard core repulsion at short distances. Other important ingredients capture the tendency of active particle to align their velocities and self-propulsion forces driving the motion. Finally, the dynamics is affected by noise.

One of the first active particle models for cell migration (Szabó et al. 2006) was constructed in analogy with flocking models originally devised to describe birds (Vicsek et al. 1995). The model considers a two dimensional overdamped equation in which the velocity of each cell is driven by a combination of an active force and the interaction with neighboring cells (Szabó et al. 2006):

$$\frac{d\mathbf{r}_i}{dt} = \mathbf{n}_i(t)v_0 + \sum_j \mathbf{f}_{ij}, \tag{5.3}$$

where v_0 is proportional to the self-propulsion force and $\mathbf{f}_{ij}$ is the force between neighboring cells due to adhesion and repulsion. The cell orientation axis $\mathbf{n}_i$ evolves according to a stochastic differential equation, parametrized by an angle θ_i

$$\frac{d\theta_i}{dt} = \xi_i(t) + \frac{1}{\tau}\arcsin\left[\hat{z}\left(\mathbf{n}_i \cdot (\mathbf{v}_i/|\mathbf{v}_i)|)\right]. \tag{5.4}$$

Finally $\xi_i(t)$ is a random uncorrelated Guassian noise. The model was used to investigate the density dependence of the cell flow patterns and was found in good agreement with experiments on one keratocytes in vitro. In particular, the authors concluded that the transition to flocking in the model is in the same universality class as in the original flocking model (Vicsek et al. 1995).

A more refined active particle model for collective cell migration was designed to study wound healing in epithelial cells (Sepúlveda et al. 2013). In this case, the equation of motion for each cell i was given by

$$\frac{d\mathbf{v}_i}{dt} = -\alpha\mathbf{v}_i + \sum_j\left[\frac{\beta}{N_i}(\mathbf{v}_j - \mathbf{v}_i) + \mathbf{f}_{ij}\right] + \sigma(\rho_i)\boldsymbol{\eta}_i + \mathbf{F}_{\mathrm{rf}}(\mathbf{x}_i) \tag{5.5}$$

where the sum is restricted to the nearest neighbors of i, α is a damping parameter, β is the velocity coupling strength and $\mathbf{f}_{ij}$ is again the interaction force. The equation of motion contains a stochastic self-propulsion force $\sigma(\rho_i)\boldsymbol{\eta}_i$, where $\boldsymbol{\eta}_i$ follows an Ornstein-Uhlenbeck process with correlation time τ:

$$\tau \frac{d\eta_i}{dt} = -\eta_i + \boldsymbol{\xi}_i, \tag{5.6}$$

$\boldsymbol{\xi}_i$ is a delta-correlated white noise $\langle \boldsymbol{\xi}_i(t) \boldsymbol{\xi}_j(t') \rangle = \delta_{ij}\delta(t - t')$. The amplitude of the noise term σ depends on the density of the neighboring cells ρ_i as

$$\sigma(\rho_i) = \sigma_0 + (\sigma_1 - \sigma_0)(1 - \rho_i/\rho_0), \tag{5.7}$$

where ρ_i and ρ_0 are the local and global particle densities, respectively. The neighbors of each cell i are found considering a circle of radius R surrounding the cell and then dividing it into 6 equal sectors. The neighbors are then defined as the cells that are closer to the cell i in each sector. The model as it is gives invasion fronts that are too diffusive when compared to the experiment. Therefore it was proposed to introduce a resistance of the medium to the invasion process (Sepúlveda et al. 2013). To this end, one can consider a set of tightly packed surface particles, that are hindering cells to enter the empty space. The interaction between a surface particle and a cell is modeled by a simple repulsive potential. Prolonged contact between particles and cells leads to the damage of the latter, allowing cells to invade. Numerical simulations of the model allows to reproduce with great accuracy the experimentally observed dynamics in an epithelial wound healing assay (see Fig. 5.5).

The model can also be used to simulate the role of leader cells (Sepúlveda et al. 2013), a subset of cells with a special phenotype that would allow them to drive the collective migration process by finding the best path and dragging the other cell with them (Khalil and Friedl 2010). In the context of the experiment illustrated in Fig. 5.5, leader cells would be associated to the formation of fingers in the front (Sepúlveda et al. 2013).

The same model was later used to study the statistical properties of front dynamics in wound healing experiments, focusing on the avalanche behavior discussed in Sect. 5.3. Numerical results show that both the avalanche distributions and the velocity distributions measured in experiments, over a wide variety of cells are well described the model. This is illustrated in Fig. 5.3b and Fig. 5.3b where the result of the simulations is compared with the experimental curves obtain from different cell lines (Chepizhko et al. 2016).

By fine tuning the parameters of Eq. 5.5 it is possible to fit quite accurately the not only the qualitative shape of velocity distributions (see Fig. 5.3d) but also the actual values velocity fluctuations and the correlation functions (Chepizhko et al. 2018). In some cases, however, a precise quantitative description can only be obtained by adding to Eq. 5.5 a self-propulsion term $F_0\hat{v}$ (Chepizhko et al. 2018), similar to the one employed in Szabó et al. (2006). Using this form of the model, it was possible to characterize experiments on epithelial cells where the induction of RAB5A leads to a dramatic fluidization of a jammed cellular monolayer (Malinverno et al. 2017). The comparison between experiments in wound healing conditions and similar experiments performed in confluent conditions shows that the appearance of a wound is able

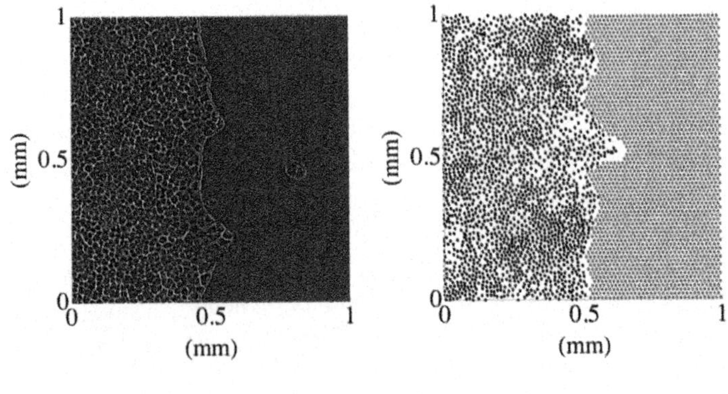

Fig. 5.5 Comparison between experiments on epithelial wound healing and simulations of an active particle model. (Image from Sepúlveda et al. 2013 (creative commons))

to induce a fluidization transition by a change of the effective parameters in Eq. 5.5. This is a marked difference with respect to ordinary soft matter systems where changes in boundary conditions do not change the internal parameters. It is instead a peculiarity of living matter where cells can change their phenotype in response to external stimuli.

5.5.2 Vertex and Voronoi Models

A different class of models of cell tissues is based on vertex models (Bi et al. 2015, 2016; Li and Sun 2014), originally developed to study foams rheology (Okuzono and Kawasaki 1995; Weaire and Kermode 1984). In those models, cells are represented by polygons whose edges and vertices are shared by neighboring cells. This representation is well suited to describe a cell monolayer or an epithelial sheet where cells are in close contact and can form tight junctions. The dynamics of the sheet is captured by equation of motions for each vertex, possibly including rules for topological changes in the edges. There is a long tradition on the application of vertex models to study tissue growth and deformation as well as cell migration.

Here, we discuss a recent development of vertex models where the moving degrees of freedom are not the vertices but the centers of the polygons (Bi et al. 2016). This case is defined as a Voronoi model, since the ensemble of polygons are part of a Voronoi tessellation of the plane. The elastic energy of each configuration composed by N polygons is similar to the one used in other vertex models and is given by

$$E = \sum_{i=1}^{N} \left[K_A (A(\mathbf{r}_i) - A_0)^2 + K_P (P(\mathbf{r}_i) - P_0)^2 \right]. \tag{5.8}$$

where $A(\mathbf{r}_i)$ and $P(\mathbf{r}_i)$ are area and perimeter of the cell i, respectively. The quadratic energy is designed to keep cell areas and perimeters close to their target values A_0 and P_0. One can thus characterize the energy by a dimensionless target shape $p_0 = P_0/\sqrt{A_0}$. Finally, K_A and K_P are the elastic moduli describing deformations of the area and the perimeter (Bi et al. 2016).

The equation of motion for each polygon is overdamped, so that the velocity is proportional to the sum of the forces given by the elastic interactions and self-propulsion

$$\frac{d\mathbf{r}_i}{dt} = \mu \mathbf{F}_i + v_0 \hat{n}_i, \tag{5.9}$$

where $\mathbf{F}_i$ is the elastic force derived from Eq. 5.8, v_0 is the self-propulsion velocity and the vector $\hat{n}_i$ indicates the polarity of the cell i. In analogy with the active particle model described in Sect. 5.5.1, the polarity is parametrized by an angle θ_i evolving as

$$\frac{d\theta_i}{dt} = \xi_i(t) \tag{5.10}$$

where $\xi_i(t)$ is again an uncorrelated Gaussian white noise.

Simulations of the self-propelled Voronoi model allow to study the behavior as a function of a few key parameters, like the self-propulsion speed v_0 and the anisotropy p_0 (Bi et al. 2016). The results are summarized in Fig. 5.6c showing the transition line between a solid-like and fluid-like phase. The phase is determined by looking at the trajectories of individual polygons. Those are confined for the solid phase and diffusive in the fluid phase. The results are similar to those discussed for the active particle models, although the system size dependence was not studied for the self-propelled Voronoi model.

While the description of the jamming transition is similar for vertex models and particle models, the latter seem more appropriate to study wound healing conditions which require particles to spread and possibly detach. Recent advances in vertex models have, however, overcome the limitations of periodic boundary conditions, allowing the study of front propagation also in this framework (Barton et al. 2017).

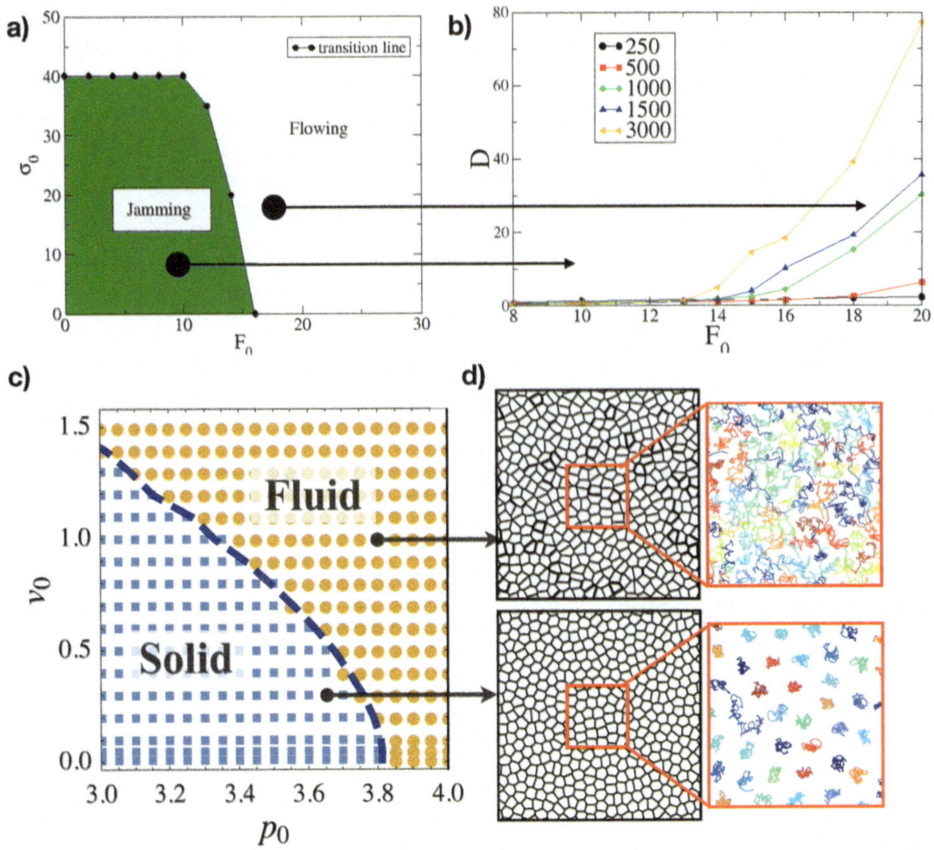

Fig. 5.6 **A comparison of active particle and vertex models.** (**a**) The phase diagram obtained from simulations of the active particle model in terms of two parameters, the noise amplitude σ_0 and the self-propulsion F_0 (Chepizhko et al. 2018). (**b**) In the jammed phase the particle self-diffusion is system size independent, while it depends on the system size for the flowing phase. (**c**, **d**) A similar phase diagram can be obtained for the self-propelled Voronoi model in terms of the self-propulsion velocity v_0 and the anisotropy parameter p_0 (Bi et al. 2016)

5.6 Conclusions

In this chapter, we have highlighted some similarities and differences between collective cell migration and the rheology of soft, but inanimate matter. Tools and ideas developed to study the physics of soft materials has been proven very useful in interpreting some properties of collective cell migration, with concept such as jamming and scaling that are being increasingly employed to describe cells. While the similarities are sometimes striking, one should always bear in mind the peculiarities of living cells that make them different from conventional soft matter and even active colloids. Cells can respond to external stimuli by changing their phenotype in a complex fashion, something that does not happen in materials. This leads to intriguing phenomena such as the boundary induced unjamming observed in confluent monolayers when a wound is produced (Chepizhko et al. 2018). These aspects may have important implications for critical biological and pathological processes such as development or tumor dissemination.

Acknowledgements We wish to thank our collaborators on the topic of collective cell migration. In particular, we would like to mention O. Chepizhko, M. C. Lionetti, C. Giampietro and G. Scita.

References

Angelini TE, Hannezo E, Trepat X, Fredberg JJ, Weitz DA (2010) Cell migration driven by cooperative substrate deformation patterns. Phys Rev Lett 104(16):168104
Angelini TE, Hannezo E, Trepat X, Marquez M, Fredberg JJ, Weitz DA (2011) Glass-like dynamics of collective cell migration. Proc Natl Acad Sci U S A 108(12):4714–4719
Banerjee S, Utuje KJC, Marchetti MC (2015) Propagating stress waves during epithelial expansion. Phys Rev Lett 114(22):228101
Barton DL, Henkes S, Weijer CJ, Sknepnek R (2017) Active vertex model for cell-resolution description of epithelial tissue mechanics. PLoS Comput Biol 13(6):e1005569
Bazellières E, Conte V, Elosegui-Artola A, Serra-Picamal X, Bintanel-Morcillo M, Roca-Cusachs P, Muñoz JJ, Sales-Pardo M, Guimerà R, Trepat X (2015) Control of cell-cell forces and collective cell dynamics by the intercellular adhesome. Nat Cell Biol 17(4):409–420
Berthier L (2014) Nonequilibrium glassy dynamics of self-propelled hard disks. Phys Rev Lett 112:220602
Dapeng Bi, Lopez JH, Schwarz JM, Lisa Manning M (2015) A density-independent rigidity transition in biological tissues. Nat Phys 11(12):1074–+
Bi D, Yang X, Marchetti MC, Manning ML (2016) Motility-driven glass and jamming transitions in biological tissues. Phys Rev X 6:021011
Bonn D, Tanase S, Abou B, Tanaka H, Meunier J (2002) Laponite: aging and shear rejuvenation of a colloidal glass. Phys Rev Lett 89(1):015701
Brugues A, Anon E, Conte V, Veldhuis JH, Gupta M, Colombelli J, Munoz JJ, Brodland GW, Ladoux B, Trepat X (2014) Forces driving epithelial wound healing. Nat Phys 10(9):683–690
Chauve P, Le Doussal P, Wiese KJ (2001) Renormalization of pinned elastic systems: how does it work beyond one loop. Phys Rev Lett 86:1785–1788
Chepizhko O, Giampietro C, Mastrapasqua E, Nourazar M, Ascagni M, Sugni M, Fascio U, Leggio L, Malinverno C, Scita G, Santucci S, Alava MJ, Zapperi S, La Porta CAM (2016) Bursts of activity in collective cell migration. Proc Natl Acad Sci U S A 113(41):11408–11413
Chepizhko O, Lionetti MC, Malinverno C, Giampietro C, Scita G, Zapperi S, La Porta CAM (2018) From jamming to collective cell migration through a boundary induced transition. Soft Matter 14(19):3774–3782
Cloitre M, Borrega R, Monti F, Leibler L (2003) Glassy dynamics and flow properties of soft colloidal pastes. Phys Rev Lett 90(6):068303
Clotet X, Ortín J, Santucci S (2014) Disorder-induced capillary bursts control intermittency in slow imbibition. Phys Rev Lett 113(7):074501
Codling EA, Plank MJ, Benhamou S (2008) Random walk models in biology. J R Soc Interface 5(25):813–834
Dieterich P, Klages R, Preuss R, Schwab A (2008) Anomalous dynamics of cell migration. Proc Natl Acad Sci U S A 105(2):459–463
Doxzen K, Vedula SRK, Leong MC, Hirata H, Gov NS, Kabla AJ, Ladoux B, Lim CT (2013) Guidance of collective cell migration by substrate geometry. Integr Biol (Camb) 5(8):1026–1035
Durian DJ, Weitz DA, Pine DJ (1991) Multiple light-scattering probes of foam structure and dynamics. Science 252(5006):686–688
Durin G, Zapperi S (2000) Scaling exponents for Barkhausen avalanches in polycrystalline and amorphous ferromagnets. Phys Rev Lett 84:4075–4078
Fily Y, Henkes S, Marchetti MC (2014) Freezing and phase separation of self-propelled disks. Soft Matter 10:2132–2140
Flenner E, Szamel G, Berthier L (2016) The nonequilibrium glassy dynamics of self-propelled particles. Soft Matter 12:7136–7149
Friedl P, Gilmour D (2009) Collective cell migration in morphogenesis, regeneration and cancer. Nat Rev Mol Cell Biol 10(7):445–457
Garcia S, Hannezo E, Elgeti J, Joanny J-F, Silberzan P, Gov NS (2015) Physics of active jamming during collective cellular motion in a monolayer. Proc Natl Acad Sci U S A 112(50):15314–15319
Gov NS (2014) Cell and matrix mechanics. CRC Press, Boca Raton, pp 219–238
Haeger A, Krause M, Wolf K, Friedl P (2014) Cell jamming: collective invasion of mesenchymal tumor cells imposed by tissue confinement. Biochim Biophys Acta 1840(8):2386–2395
Haga H, Irahara C, Kobayashi R, Nakagaki T, Kawabata K (2005) Collective movement of epithelial cells on a collagen gel substrate. Biophys J 88(3):2250–2256
Henkes S, Fily Y, Marchetti MC (2011) Active jamming: self-propelled soft particles at high density. Phys Rev E 84:040301
Ilina O, Friedl P (2009) Mechanisms of collective cell migration at a glance. J Cell Sci 122(Pt 18):3203–3208
Khalil AA, Friedl P (2010) Determinants of leader cells in collective cell migration. Integr Biol (Camb) 2(11–12):568–574
Koch TM, Münster S, Bonakdar N, Butler JP, Fabry B (2012) 3D traction forces in cancer cell invasion. PLoS One 7(3):e33476
La Porta CAM, Zapperi S (2017) The physics of cancer. Cambridge University Press, Cambridge
Lange JR, Fabry B (2013) Cell and tissue mechanics in cell migration. Exp Cell Res 319(16):2418–2423
Le Doussal P, Wiese KJ (2009) Size distributions of shocks and static avalanches from the functional renormalization group. Phys Rev E 79:051106
Leschhorn H, Nattermann T, Stepanow S, Tang LH (1997) Driven interface depinning in a disordered medium. Ann Physik 6:1–34
Li B, Sun SX (2014) Coherent motions in confluent cell monolayer sheets. Biophys J 107(7):1532–1541
Li L, Nørrelykke SF, Cox EC (2008) Persistent cell motion in the absence of external signals: a search strategy for eukaryotic cells. PLoS One 3(5):e2093

Liao Q, Xu N (2018) Criticality of the zero-temperature jamming transition probed by self-propelled particles. Soft Matter 14:853–860

Liu AJ, Nagel SR, Langer JS (2010) The jamming transition and the marginally jammed solid. Annu Rev Condens Matter Phys 1:347–369

Malinverno C, Corallino S, Giavazzi F, Bergert M, Li Q, Leoni M, Disanza A, Frittoli E, Oldani A, Martini E, Lendenmann T, Deflorian G, Beznoussenko GV, Poulikakos D, Haur OK, Uroz M, Trepat X, Parazzoli D, Maiuri P, Yu W, Ferrari A, Cerbino R, Scita G (2017) Endocytic reawakening of motility in jammed epithelia. Nat Mater 16(5):587–596

Maloy KJ, Santucci S, Schmittbuhl J, Toussaint R (2006) Local waiting time fluctuations along a randomly pinned crack front. Phys Rev Lett 96:045501

Mandal R, Bhuyan PJ, Rao M, Dasgupta C (2016) Active fluidization in dense glassy systems. Soft Matter 12:6268–6276

Marchetti MC, Banerjee S (2019) Continuum models of collective cell migration. In: La Porta CAM, Zapperi S (eds) Cell migrations: causes and functions. Springer, Cham

Mason TG, Bibette J, Weitz DA (1996) Yielding and flow of monodisperse emulsions. J Colloid Interface Sci 179(2):439–448

Metzner C, Mark C, Steinwachs J, Lautscham L, Stadler F, Fabry B (2015) Superstatistical analysis and modelling of heterogeneous random walks. Nat Commun 6:7516

Narayan O, Fisher DS (1992) Critical behavior of sliding charge-density waves in 4- epsilon dimensions. Phys Rev B 46:11520

Ng MR, Besser A, Danuser G, Brugge JS (2012) Substrate stiffness regulates cadherin-dependent collective migration through myosin-ii contractility. J Cell Biol 199(3):545–563

Okuzono T, Kawasaki K (1995) Intermittent flow behavior of random foams: a computer experiment on foam rheology. Phys Rev E 51:1246–1253

Park J-A, Kim JH, Bi D, Mitchel JA, Qazvini NT, Tantisira K, Park CY, McGill M, Kim S-H, Gweon B, Notbohm J, Steward R Jr, Burger S, Randell SH, Kho AT, Tambe DT, Hardin C, Shore SA, Israel E, Weitz DA, Tschumperlin DJ, Henske EP, Weiss ST, Manning ML, Butler JP, Drazen JM, Fredberg JJ (2015) Unjamming and cell shape in the asthmatic airway epithelium. Nat Mater 14:1040–1048

Potdar AA, Jeon J, Weaver AM, Quaranta V, Cummings PT (2010) Human mammary epithelial cells exhibit a bimodal correlated random walk pattern. PLoS One 5(3):e9636

Poujade M, Grasland-Mongrain E, Hertzog A, Jouanneau J, Chavrier P, Ladoux B, Buguin A, Silberzan P (2007) Collective migration of an epithelial monolayer in response to a model wound. Proc Natl Acad Sci U S A 104(41):15988–15993

Ramaswamy S (2010) The mechanics and statistics of active matter. Annu Rev Condens Matter Phys 1(1): 323–345

Rørth P (2009) Collective cell migration. Annu Rev Cell Dev Biol 25:407–429

Rosso A, Le Doussal P, Jörg Wiese K (2009) Avalanche-size distribution at the depinning transition: a numerical test of the theory. Phys Rev B 80:144204

Röttgermann PJF, Alberola AP, Rädler JO (2014) Cellular self-organization on micro-structured surfaces. Soft Matter 10(14):2397–2404

Sacks MS, Sun W (2003) Multiaxial mechanical behavior of biological materials. Annu Rev Biomed Eng 5:251–284

Saez A, Ghibaudo M, Buguin A, Silberzan P, Ladoux B (2007) Rigidity-driven growth and migration of epithelial cells on microstructured anisotropic substrates. Proc Natl Acad Sci U S A 104(20):8281–8286

Segrè PN, Prasad V, Schofield AB, Weitz DA (2001) Glasslike kinetic arrest at the colloidal-gelation transition. Phys Rev Lett 86(26 Pt 1):6042–6045

Sepúlveda N, Petitjean L, Cochet O, Grasland-Mongrain E, Silberzan P, Hakim V (2013) Collective cell motion in an epithelial sheet can be quantitatively described by a stochastic interacting particle model. PLoS Comput Biol 9(3):e1002944

Serra-Picamal X, Conte V, Vincent R, Anon E, Tambe DT, Bazellieres E, Butler JP, Fredberg JJ, Trepat X (2012) Mechanical waves during tissue expansion. Nat Phys 8(8):628–634

Stokes CL, Lauffenburger DA, Williams SK (1991) Migration of individual microvessel endothelial cells: stochastic model and parameter measurement. J Cell Sci 99(Pt 2):419–430

Szabó B, Szöllösi GJ, Gönci B, Jurányi ZS, Selmeczi D, Vicsek T (2006) Phase transition in the collective migration of tissue cells: experiment and model. Phys Rev E 74:061908

Szamel G (2016) Theory for the dynamics of dense systems of athermal self-propelled particles. Phys Rev E 93:012603

Tallakstad KT, Toussaint R, Santucci S, Schmittbuhl J, Maloy KJ (2011) Local dynamics of a randomly pinned crack front during creep and forced propagation: an experimental study. Phys Rev E Stat Nonlin Soft Matter Phys 83(4 Pt 2):046108

Tambe DT, Hardin CC, Angelini TE, Rajendran K, Park CY, Serra-Picamal X, Zhou EH, Zaman MH, Butler JP, Weitz DA, Fredberg JJ, Trepat X (2011) Collective cell guidance by cooperative intercellular forces. Nat Mater 10(6):469–475

Vedula SRK, Ravasio A, Lim CT, Ladoux B (2013) Collective cell migration: a mechanistic perspective. Physiology (Bethesda) 28(6):370–379

Vicsek T, Czirók A, Ben-Jacob E, Cohen I, Shochet O (1995) Novel type of phase transition in a system of self-driven particles. Phys Rev Lett 75:1226–1229

Weaire D, Kermode JP (1984) Computer simulation of a two-dimensional soap froth ii. analysis of results. Philos Mag B 50(3):379–395

Wu P-H, Giri A, Sun SX, Wirtz D (2014) Three-dimensional cell migration does not follow a random walk. Proc Natl Acad Sci U S A 111(11):3949–3954

Cell Migration in Microfluidic Devices: Invadosomes Formation in Confined Environments

6

Pei-Yin Chi, Pirjo Spuul, Fan-Gang Tseng, Elisabeth Genot, Chia-Fu Chou, and Alessandro Taloni

Abstract

The last 20 years have seen the blooming of microfluidics technologies applied to biological sciences. Microfluidics provides effective tools for biological analysis, allowing the experimentalists to extend their playground to single cells and single molecules, with high throughput and resolution which were inconceivable few decades ago. In particular, microfluidic devices are profoundly changing the conventional way of studying the cell motility and cell migratory dynamics. In this chapter we will furnish a comprehensive view of the advancements made in the research domain of confinement-induced cell migration, thanks to the use of microfluidic devices. The chapter is subdivided in three parts. Each section will be addressing one of the fundamental questions that the microfluidic technology is contributing to unravel: (i) *where* cell migration takes place, (ii) *why* cells migrate and, (iii) *how* the cells migrate. The first introductory part is devoted to a thumbnail, and partially historical, description of microfluidics

Pei-Yin Chi and Pirjo Spuul contributed equally with all other contributors.

P. -Y. Chi
Department of Engineering and System Science, National Tsing Hua University, Hsinchu, Taiwan, Republic of China

Nano Science and Technology Program, Taiwan International Graduate Program, Academia Sinica, Taipei, Taiwan, Republic of China

Institute of Physics, Academia Sinica, Taipei, Taiwan, Republic of China

P. Spuul
Department of Chemistry and Biotechnology, Division of Gene Technology, Tallinn University of Technology, Tallinn, Estonia

F.-G. Tseng
Department of Engineering and System Science, National Tsing Hua University, Hsinchu, Taiwan, Republic of China

Frontier Research Center on Fundamental and Applied Sciences of Matters, National Tsing Hua University, Hsinchu, Taiwan, Republic of China

Research Center for Applied Sciences, Academia Sinica, Taipei, Taiwan, Republic of China

E. Genot (✉)
Centre de Recherche Cardio-Thoracique de Bordeaux (INSERM U1045), Université de Bordeaux, Bordeaux, France
e-mail: e.genot@iecb.u-bordeaux.fr

C.-F. Chou (✉)
Institute of Physics, Academia Sinica, Taipei, Taiwan, Republic of China

Research Center for Applied Sciences, Academia Sinica, Taipei, Taiwan, Republic of China

Genomics Research Center and Research Center for Applied Sciences, Academia Sinica, Taipei, Taiwan, Republic of China
e-mail: cfchou@phys.sinica.edu.tw

A. Taloni (✉)
Institute for Complex Systems, Consiglio Nazionale delle Ricerche, Roma, Italy

C. A. M. La Porta, S. Zapperi (eds.), *Cell Migrations: Causes and Functions*, Advances in Experimental Medicine and Biology 1146,
https://doi.org/10.1007/978-3-030-17593-1_6

and its impact in biological sciences. Stress will be put on two aspects of the devices fabrication process, which are crucial for biological applications: materials used and coating methods. The second paragraph concerns the cell migration induced by environmental cues: chemical, leading to chemotaxis, mechanical, at the basis of mechanotaxis, and electrical, which induces electrotaxis. Each of them will be addressed separately, highlighting the fundamental role of microfluidics in providing the well-controlled experimental conditions where cell migration can be induced, investigated and ultimately understood. The third part of the chapter is entirely dedicated to *how* the cells move in confined environments. Invadosomes (the joint name for podosomes and invadopodia) are cell protrusion that contribute actively to cell migration or invasion. The formation of invadosomes under confinement is a research topic that only recently has caught the attention of the scientific community: microfluidic design is helping shaping the future direction of this emerging field of research.

Keywords
Microfluidics · Coating · Cell migration · Chemotaxis · Mechanotaxis · Haptotaxis · Durotaxis · Plithotaxis · Electrotaxis · Invadosomes · Podosomes · Invadopodia · Confinement · Microenvironment

6.1 Microfluidic Devices in Biological Applications

Microfluidics deploys microfabricated structures with dimensions going from 1 to 500 μm and a volume capacity between 10^{-9} and 10^{-15} l. Microfluidic devices consist of a multiplicity of common components: negative features refer to empty spaces such as chambers, wells and microchannels; positive components are filled volumes of solid material like membranes, pillars and beams, among others. Initially thought for liquid handling (Tabeling 2005; Kirby 2010; Chang and Yeo 2009; Manz et al. 1990), microfluidics over time has displayed its enormous potential in life sciences applications (Gravesen et al. 1993; Mark et al. 2010; Wheeler et al. 2003; Zare and Kim 2010; Yin and Marshall 2012; Sriram et al. 2014). Indeed, exploiting the large heterogeneity of components, their versatility in terms of shape, rugosity and materials, researchers could on one hand broaden their field of investigation, and on the other ease the observations, by controlling the complete cellular environment. These components, while integrated together, gave rise to the trailblazing idea of "lab on a chip", in the sense that an entire lab could be fit into a single microfluidic device, inasmuch a microelectronic circuit can be thought as a computer on a chip. The true turning point was the rapid expansion of soft-lithography in the decade of 1995–2005: since then, a simple and low-cost technology for fabricating devices that combine channels and other microelements of the cell dimensions, became suddenly within reach of many experimental groups worldwide (Kim et al. 2008). Microfluidic channels, indeed, offer unquestionable advantages such as large surface-to-volume ratio, small overall volumes and laminar flow, to mention few.

In most cases, the early biologists' enthusiasm for microfluidics stem from the practical experimental advantages that this technology was offering, rather than being driven by the possibility of observing new physics at the microscale. Microfluidic devices, indeed, are physically and chemically well-controlled environments. At the same time, microdevices allowed a systematic probe of the scaling of physical forces at the microscale: while the physics laws stay unaltered as compared to macroscopic systems, the scale factor can play a crucial role in microscopic environments, leading different forces to prevail according to the systems design (Velve-Casquillas et al. 2010; Yeo et al. 2011).

To provide a quick example, one of the first successes of microfluidics in biology has been the cells culture medium fine control, in order to mimic accurately the chemical gradients occurring in many biological processes (Li and Francis 2011; Sackmann et al. 2014). Cells respond to chemical gradients which can be confined to a region much smaller than their size. Thus, investigating their response to gradients, like the study of the cells migratory activity responsiveness to chemokines (Keenan and Folch 2008), requires an extremely high spatial control of the media concentrations. Opposite to macroscopic gradient generators, which offer a rather the scarse spatiotemporal resolution, microfluidic devices can create a large spectrum of biochemical gradients with highly controlled distribution in space and time and subcellular resolution: time invariant gradients, fast response dynamic gradients, continuous or discrete gradients, subcellular resolution gradients, are just few examples (Irimia et al. 2006).

Although the early microfluidic research was deeply rooted on "analytical chemistry," later the scientific interest switched toward cell-based systems and toward the biochemical experimentation/analysis. This transition arose quite naturally if one considers the fundamental breakthrough that microfluidic technologies have enabled, the possibility to handle with an unprecedented accuracy single-cells and single-molecules. For instance, the estimation of forces at the microscale, the understanding of the role played by mechanical cues and confining microenvironments on cells functionalities, have largely stimulated the use of microfluidics in mechanobiology. However, due to the intrinsic dynamical nature of cells, the comparison between quantities measured in different labs, even using the same experimental procedures, turns out to be very problematic, as in the case of the estimation of small forces. Nevertheless, the heterogeneity that one can find in real living systems is even higher, and (mechano)biologists may gain fundamental insights from appropriately designed experiments and protocols, in order to assess the difference between theoretical models and experimental outcomes (Yeh et al. 2012).

Microfluidics has also weaknesses. For example, microfluidics cannot be thought as valuable tool in those applications in which fast flow homogenization embodies a crucial aspect, as it only produces slow diffusion-driven mixing. Yet, this limitation can be rectified by having a recourse to diverse integrated mixers. Moreover, the changes in scaling laws may be the cause of considerable problems while attempting to adapt biological protocols to fit experiments in microdevices. By instance, changing the osmolarity, the permeability to water vapor of the elastomers used in the microfabrication processes may lead to media drying. Thus, one has to be extremely careful while comparing data arising from macroscopic experiments and data from microsystems (Young and Beebe 2010). A thorough and critical discussion about the microfluidic toolset and its enormous impact on the chemical and biological sciences, as well as on the future opportunities that it will provide and challenges that it will be facing, can be found in (Chiu et al. 2017).

6.1.1 Materials

At the dawn of macrofabrication era, structures were mainly made of silicon and glass. Since then, this technology has impressively expanded including an enormous variety of components, equipments and materials. These materials can be summarily grouped into three classes: inorganic, polymeric and paper (Nge et al. 2013). Nowadays, the inorganic materials domain goes far beyond glass and silicon, encompassing substrata such as vitroceramics and ceramics co-fired with metal conductors. Thermoplastics and elastomers alone, instead, complete the class of polymer-based materials. Paper microfluidics is a substantially different and emergent technology, for which we refer the reader to specialized reviews (Martinez et al. 2009).

The microfluidic device material initially employed in biological applications was mainly silicon. The nonspecific proteins adsorption typical of silicon substrates can be seriously diminished, and the cellular viability improved, through silanol based surface chemical exposition (Li et al. 2011). However, since silicon is transparent to infrared but opaque to visible light, usual techniques as fluorescence identification and fluid imaging for complex biological structures, may result to be particularly arduous in silicon-made chips (Evstrapov 2017; Sriram et al. 2017). For this reason essentially, the focus shifted to glass and, later, included polymer substrates. It must be said that this problem, although serious, can be partially circumvented if a transparent material, like polymer or glass, is bound to silicon in a hybrid system. Devices of this kind are droplet-based microfluidic systems for the polymerase chain reaction (PCR), or highly aligned silicon nanowire (SiNW) clusters for label-free cardiac biomarker detection. These hybrid class of microdevices have led to the recent renaissance of Si-based devices in biology.

After silicon, glass came on the scene, emerging as a suitable substrate material for a time. In this case, structures are obtained by wet or dry etching techniques into the glass surface (Sriram et al. 2017). The poor background fluorescence requires, as in the case of silicon, silanol based modification chemistries. Glass is not gas permeable (Evstrapov 2017) and, most importantly, is a viable substrate. The weakness of glass microdevices is that the substance itself is rather expensive.

The widespread use of elastomers as the preferential material in chips microfabrication is mainly due to two factors: the low cost and the compatibility with biological samples. Among elastomers, PDMS (Polydimethylsiloxane) holds a prominent position. Commonly known as silicone rubber, it is widely used in lubricating oils as well as in the food industry (it is the additive named E900) and cosmetics industry. Below we list and briefly discuss some of the properties that make PDMS one of the best materials for microdevices:

- **Transparency and other properties:** PDMS is optically transparent (UV transparency > 220 nm) (Nge et al. 2013). Therefore, micro-channels, micro-compartments and their content can be visualized directly as glass or silica substrate (Liao and Chou 2012). Moreover, PDMS is a non-toxic material with negative surface charge. Electrically it is an insulator (breakdown voltage, 2×10^7 V/m) (McDonald and Whitesides 2002; Gu et al. 2007)
- **Elasticity/Stiffness**: The elasticity of PDMS makes this material particularly suited for various applications. Stiffness has a role in orienting cell division, maintaining tissue boundaries, driving differentiation and cell viability (Handorf et al. 2015; Su et al. 2015). For instance, seeding of fibroblasts on PDMS without any ECM coating produced 45% reduction in the cell viability (Park et al. 2010a, b). Importantly, the substrate stiffness has been proven to have a direct impact on cell migration, as discussed in more details in the Sect. 6.2.2.2. From this perspective, the fact that PDMS elasticity can be "tuned", is of fundamental importance. The stiffness of the PDMS increases proportionally to the increase of curing agent as follows: 20:1, 10:1, 5:1, resulting in a Young's modulus (E) equal to 280 kPa, 580 kPa, and 1000 kPa respectively (Park et al. 2010a, b). Moreover, a stiffness ranging from E = 5 kPa up to 1.72 Mpa, has been achieved using different ratios of commercially available cross-linking agents, Sylgard 527 and Sylgard 184 (Palchesko et al. 2012). Having such a wide range of tunable elasticity is extremely important for studying cell response to substrate mechanics (Nemir and West 2010). Such a large range of attainable elasticity, indeed, covers most of the stiffness of native tissues and organs (gray matter~1.4 kPa (Budday et al. 2015), cartilage ~ 2.6 Mpa (Stoltz et al. 2004)), as discussed in (Liu et al. 2015).
- **Cost**: As mentioned before, PDMS is perhaps the most widely used microfabrication material in academic facilities. Indeed, it is much less expensive than other microfluidic materials, and its rapid prototyping makes it an excellent choice for mass production devices. It is very easy to handle and the well-defined topography of PDMS replica on nanometer-scale (less than 100 nm) is also a standard technique nowadays (Kim et al. 2002).
- **Permeability**: PDMS is gas permeable, which is extremely useful in a vast series of devices. In particular PDMS is O_2 permeable, which is advantageous in biological application, like cell culture or migration. The estimated diffusion coefficient for O_2 in PDMS at 300 K is 16×10^6 (cm^2/s) by fractional free volume calculated through the lattice-search method (Charati and Stern 1998).

Besides these properties, PDMS incorporates several features that are fundamental in biological applications. First, PDMS is a biocompatible material (Sherman et al. 1999). Methods to improve its biocompatibility are quite cheap and simple, as to boil the PDMS for 5–60 min (Park et al. 2010a, b), treating it with O_2 plasma, or to undergo ECM coating (see Sect. 6.1.2). Secondly, PDMS is a solvent compatible elastomer. The stability of the PDMS materials in solvent is truly important, since PDMS is used in the production of devices loaded with solutions. For a thorough discussion on the compatibility of PDMS to different solutions or organic solvents, we refer the reader to the review (Lee et al. 2003). In particular, for biological applications the swelling ratio is an important parameter to take into account, since devices are often treated with alcohol for sterilization, or water (cell culture medium). The swelling ratio is defined by $S = D/D_0$, where D is the length of PDMS in the solvent and D_0 is the length of the dry PDMS. The swelling ratio of water and alcohol is 1.00, and 1.04 respectively, meaning that PDMS filled with those solutions will not exhibit large structure deformation.

However, the use of PDMS also has disadvantages. PDMS is an oligomer characterized by a low molecular weight, which makes it possible to percolate into the medium solution, thus affecting the viability of cells culture. Due to its hydrophobic nature, PDMS can be easily permeated by hydrophobic molecules and often subject to non-specific proteins adsorption. As a matter of fact, the contact angle in the PDMS-based microfluidics increases to 115° after 6 h of exposure to the ambient air (Leichlé et al. 2012; Tan et al. 2010). Chemical modification of PDMS or its sister compounds can often address these issues: surface can be made hydrophilic by exposing it to air or oxygen plasma, introducing silanol groups (Si-OH) instead of methyl (Si-CH_3) groups. After the plasma treatment, PDMS can rapidly regain hydrophobic groups if it is prolongedly exposed to air. On the other side, its hydrophilic state can be protracted indefinitely by keeping the surface in water or other polar organic solvents. In any case, albeit PDMS is considered maybe the best material for prototyping uses and the most common substrate in academic labs, it is outclassed by other materials for commercial microfluidic end products.

Among elastomers, (TPE) is obtained by the copolimerization of polyester and styrene at high temperature (UV and/or heat). Using commercialized thermoset polyester, microfluidic devices are developed as an alternative to PDMS, as in the case of the multichannel hybrid chip developed for protein immobilization (Brassard et al. 2011). If compared to PDMS, polyfluoropolyether diol methacrylate (PFPE-DMA) does not swell while put in contact with organic solvents. At the same time it is stiffer than PDMS (Young's modulus nearly 10 times higher), and it offers a very high precision in the molding process, reaching a resolution of 50 nm.

In general, thermoplastics are transparent to visible light and extremely versatile, resistant to small molecules permeation, and much more rigid than elastomers. Polystyrene (PS) is usually adopted for fabrication of biocompatible devices, as it enjoys many of the properties which are essential for cell culture and analysis (Berthier et al. 2012). Poly(methyl methacrylate) (PMMA), also known under the trade names of Plexiglas and Lucite among several others, is obtained from methyl methacrylate, a liquid monomer, by polimerization (polymethyl methacrylate is indeed a solid polymer). PMMA offers several important advantages, ranging from the compatibility with biological samples, to the impossibility of being gas permeable, to the low temperature microfabrication (~100 °C). In analogy to PDMS, several acrylic monomers undergo the polimerization process on a mold, resulting in micropatterned substrates with interface characteristics that can be easily adjusted to the experimental scope. As PDMS, nonspecific proteins and cells adsorption constitutes one of the most serious disadvantages for PMMA (Kim et al. 2006), which can be partially reduced by surface incorporation of poly(ethylene glycol) (PEG) or poly(ethylene glycol) diacrylate (PEGDA) (Rogers et al. 2011). Teflon microdevices are free of this problem, having low or absent protein nonspecific adsorption if compared to PDMS or PS. Moreover teflon exhibited a very high biocompatibility (more than 5 days) and remarkable gas permeability, as it was shown in the case of HepG2 cells (Ren et al. 2011). Cyclic-olefin copolymer (COC), finally, is transparent to visible light and a versatile material, compatible with a wide range of acqueous solutions and solvents. COC devices have been shown to provide an extremely fast detection of *E. Coli, S. aureus* and *S. epidermidis* (<20 min). (Peng et al. 2010). Finally, materials such as polyethylene glycol (PEG) (Liu et al. 2015), poly(lactic-co-glycolic acid) (PLGA) (McUsic et al. 2012; Chaurey et al. 2012), or photoresist such as SU-8 (Salomon et al. 2011) have been largely used for the fabrication of biocompatible devices.

In the last two decades, as the importance of the mechanical and environmental cues on cell behavior was progressively unveiling, microfluidics devices and microfabrication technologies provided the suitable tools for the systematic study the cells-substrate interactions. Substrates exhibiting the range of elasticity typical of many soft tissues found throughout the body (0.1 kPa < E < 100 kPa), are made with Polyacrylamide (PA). Tissue-specific stiffness has been also obtained in other hydrogels chips. In this case hydrogels can be synthetic, like polyethylene glycol, or natural, based on hyaluronan, dextrane and among many others. The variability of stiffness in the body however, is even greater and many tissues, like basal membranes or those formed by collagen, enzymes, and glycoproteins, are much stiffer than any hydrogels substrate.

6.1.2 ECM Coating

Cells binds to the extracellular matrix (ECM) in vivo, in particular their anchorage is provided by carbohydrate moieties and proteins in it (Alberts et al. 2002). The ECM composition depends on tissues and organs, as it is the product principally of connective tissue but it can be also produced by other cells, like epithelial. Besides providing the right structure supporting cells binding, ECM contains intrinsic biochemical and mechanical cues for tissues segregation and intracellular communication, needed for the cells migratory dynamics regulation. Differentiated cells no longer show distinctive attributes, if they are isolated from ECM and cultured on substrates without supportive coating or feeder layer. Nevertheless ECM coating has only recently begun to be considered a key aspect of microfluidic design for biological purposes. Indeed, although the overwhelming majority of the current research is directed to the investigation of the influence of soluble factors on cell growth and migration (Duffy et al. 2008; Yoshida et al. 2008), numerous studies have undoubtedly pointed at the relevant role that tissue-specific ECM coatings play in maintaining cultured cells phenotypes and functionalities.

Tissue-specific biocompatible substrata for culture dishes and microfluidic devices are frequently formed through commercially purchasable ECM individual components (McClelland et al. 2008). Matrix elements such as collagen fibers, or glycoproteins such as fibronectin or laminin, are largely employed in cell culture because of their capability of providing the correct substratum anchorage to cells and to prolong their survival. At the same time, it has undoubtedly demonstrated that these components have a deep effect in the maintenance of several vital functions, like cells migration, differentiation and locomotion (Zhang et al. 2009).

To create biological compatible microenvironment coating in PDMS based microfluidic devices, it is highly recommendable to treat the device with oxygen plasma prior to loading the ECM solution. Oxygen plasma indeed, improves considerably the chip's uniform hydrophilicity and cleanness. Without oxygen plasma treatment, the ECM solution has to be introduced by brute force because of the hydrophobicity building up at the microscale. Furthermore, a hydrophobic–hydrophilic dichotomy may develop in PDMS microchannels bonded with a glass cover, as glass is a hydrophilic material opposite to PDMS. Therefore, the uniformity of the ensuing ECM deposition can be considerably affected. However, deposition of ECM solution directly on the surface is possible for open devices such as cell culture dishes, glass substrates or big loading wells, if immersed in the ECM solution long enough such as 1.5 h. or overnight at 25 or 37 °C. A typical PDMS-based microfluidic device, coated with ECM solutions is show in Fig. 6.1.

ECM coating is extremely useful also for quantifying the effect of mechanical cues on cellular response. As it will be amply discussed in the Sect. 6.2.2, many types of adherent cells exert contractile forces on the anchoring substratum to gain useful informations about the surrounding mechanical microenvironment. In this sense, ECM acts as a passive mechanical medium, since cells can actively reshape the ECM structure that they are in contact to. But cells can also infer mechanical clues from the surrounding ECM, while this is actively applying a force onto them: this typically happens when tissue components are sheared, elongated or compressed. Yet, ECM can function as a medium to convey mechanical informations in a small-to-long range fashion. As a matter of fact cells which are not in direct contact, may interact one to another through traction-induced ECM spatial inhomogeneities.

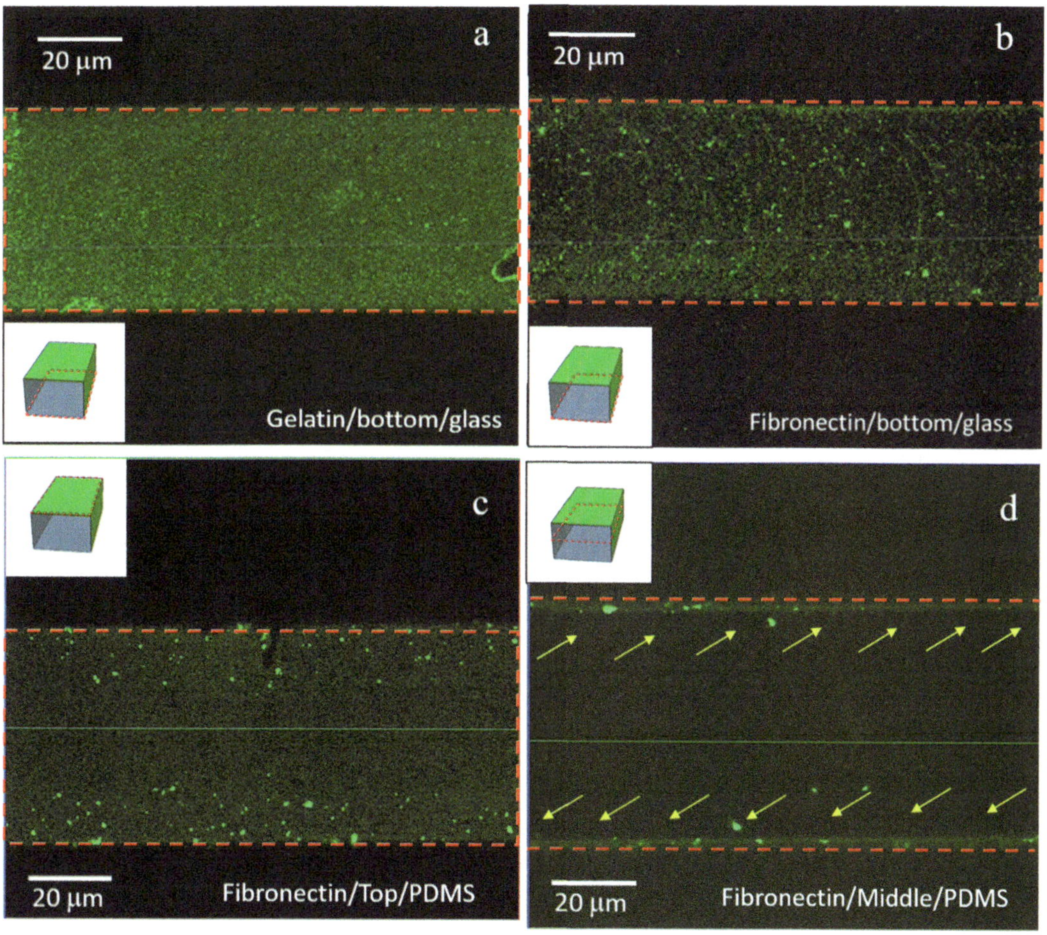

Fig. 6.1 ECM coating of a microchannel. A microchannel of height 3 μm, length 250 μm and width 50 μm, is coated with gelatin and fibronectin at the bottom glass surface (panels **a** and **b**). (**c**) Fibronectin coating on the PDMS top surface. (**d**) An image of the middle plane (dashed red lines) shows that only the channel PDMS walls are coated by fibronectin, pointed by yellow arrows

Many kind of cells usually attach primarily to the ECM rather than stick to other cells. This offers an enormous practical advantage while setting up an experimental protocol that can mimick as close as possible the in vivo conditions: cells can be put within or on an ECM-enriched substrate, and their mechanical behaviour can be acquired as an output. Therefore, in order to reveal the influence of different tissue-specific ECM components on cell functionalities in confined environments, coating microfluidic devices is mandatory. For instance, bovine aortic endothelial (BAE) cells have shown different migration speed and different percentage of cells forming podosomes in the fibronectin and non-fibronectin coated microfluidic channels (Spuul et al. 2016a). NIH 3 T3 fibroblasts are known to migrate towards stiffer substrate (see Sect. 6.2.2.2). However, this happens on fibronectin coated mechanical gradients but not on laminin ones. Interestingly, combining the two matrix components does not restore the durotaxis, highlighting the complexity of the cellular response to microenvironment (Hartman et al. 2017).

6.2 Cell Migration in Microfluidic Devices

Cell motility is the capacity of cells to translocate onto a solid substratum. This may occur within complex, multicellular organisms or in non-live environments (Ridley et al. 2003; Alt et al. 2012). Cells migration is a complex biophysical phenomenon arising as a consequence of a multiplicity of reasons, ranging from their forage, to a series of morphogenetic events requiring precursors activity, such as the need to generate biostructures even at distant sites, or the proximity of environmental cues which trigger the cells motion to attain a specific goal. These directional cues can be classified into three types: chemotactic, i.e. induced by chemoattractants, mechanotactic (caused by the mechanical properties of the surroundings) and electrotactic (induced by electric fields), or combinations of any of these (Lauffenburger and Horwitz 1996).

6.2.1 Chemotaxis

For most eukaryotic cell migration pathways *in vivo*, the first step in cell migration involves the sensing of a gradient. This directional motion in response to chemical gradients is generally referred to as chemotaxis (Van Haastert and Devreotes 2004): chemotactic stimuli, being chemoattractants or morphogens, can be perceived by the cell through the use of 7-transmembrane heterotrimeric G-protein-coupled receptors (Hamm 1998). The G-protein activation initiates a series of intracellular signaling cascade in terms of the massive production of second messenger molecules, small molecules that coordinate the signaling pathways, which eventually culminate in the actin filaments polymerization (Hepler and Gilman 1992). Indeed, the second step after signaling is cell polarization i.e., the cells propensity to express a stable front and rear during migration, involving reorganization of their actin cytoskeleton. Front and rear are often referred to as leading and trailing edges and they are characterized by very different biomechanical and morphological structures. Leading edge exhibites a vivid and protracted actin polimeryzation which produces protrusive structures, favoring the substratum adhesion. Stable bundles instead form at the cell trailing edge, and it is not entirely clear how this stability can promote the fast release and ensuing disassembly of focal adhesion sites. The nucleus and microtubules are contained in the cell central body. Third, new focal contacts are formed and the cortex contracts, dragging the cell body forward. This process continues until the cell reaches the chemokine source. Once it is surrounded by a high concentration of the chemokine and the cell no longer senses a gradient, migration will cease.

The need of having highly controlled chemical concentration gradients, led to the production of microscale gradient generators(μGGs) (Irimia et al. 2006; Keenan and Folch 2008; Kim et al. 2010; Teerapanich et al. 2018). Compared with conventional cell migration assays, flow-based microfluidics has revolutioned the way chemotaxis studies were carried out, providing a simple and reproducible experimental framework for the quantitative analysis of the migration dynamics under highly-controlled gradient conditions (Englert et al. 2009). Following an initial development where migratory cell types were investigated (e.g. bacteria and neutrophils), new studies have focused on neurons, stem cells, cancer and tissue cells among several others. One of the striking advantages offered by gradient microgenerators is the precise and flexible space-time control of chemical gradients: this has been largely exploited to address the cell responsive dynamics to temporal perturbations or complex spatial geometries (Irimia 2010; Li and Francis 2011). Flow-based devices, however, do not complete the moltitude of the gradient microgenerator devices, as simpler and more realistic microfluidic gradient systems, such as flow-free devices, were recently manufactured to address the experimental needs (Kim et al. 2010; Wu et al. 2013). By instance, the need of finely controlling chemical diffusion and increase the hydraulic resistance, has required the insertion of physical barriers such as microfabricated membranes or gel grids into the device.

6.2.2 Mechanotaxis

Cells directional motility is also influenced by the external forces as well as by the mechanical features of the surrounding environment (Lange and Fabry 2013; Roca-Cusachs et al. 2013). This directed movement goes under the name of mechanotaxis and corresponds to the cell dynamical response to environmental cues such as fluidic shear stress, substrate stiffness gradients and stress anisotropy among others.

Living cells can sense mechanical forces or deformation by i.e., transmembrane proteins or cell structures aimed at converting the external stress into an internal biochemical output. The group of activities performed by mechanosensors is called mechanosensing, and it starts with the transmembrane proteins conformational changes caused by an external mechanical perturbation, followed by the alteration of the normal interplay with the internal agonists. Mechanosensing is inherently connected and sometimes overlapping to mechanotransduction. The latter is defined as the set of molecular processes involved in the translation of the mechanical stimuli into biochemical signals (Alenghat and Ingber 2002). Usually, the activation of mechanotransduction pathways by the mechanosensed external force coincides with a second messengers massive production such as cAMP. In addition to proteins conformational changes, mechanotransduction is induced by mechanosensitive or stretch-gated ion channels deformation, or by the modification of the equilibrium configuration of G-protein-coupled receptors and of several more proteins binding the ECM via focal adhesions. Yet, the mechanosensing machinery and the activation of the ensuing transduction pathways, is sensitive to the type of stress exerted on the cell surface. Shear stress indeed is responsible for the tensile- and bending-induced conformational changes of specific transmembrane proteins, which, in turn, activate the MAPK pathway (Vogel and Sheetz 2006). During mechanotransduction stresses and biochemical signals propagate across the cell, while mechanosensing occurs mostly at the cell periphery, precisely at the focal adhesion junctions, as it has been recently shown. Some of the conformational changes, however, involve cell inner structures such as the cytoskeleton and the nucleus. The phase after mechanotransduction is the mechanoresponse, i.e. the way the cell adjusts to the external stimuli. This includes the modification of gene expression, the capability of reshaping, stiffening or softening, adapting Ca intake, and the assembly or disassembly of cell-matrix adhesions (Lange and Fabry 2013). Mechanoresponse ultimate effect is to generate an internal force imbalance, such to push the cell in a certain direction. Such dynamical instability is the result of the interaction and polarization of biochemical and mechanical elements i.e., actin-driven protrusions at the leading edge and myosin-induced contractile forces at the trailing edge. (Parsons et al. 2010).

Mechanotaxis can be subdivided into three different "taxis" i.e., different ways of cell locomotion, according to the different mechanical stimuli to which the cell has to face: haptotaxis, durotaxis and plithotaxis. In the following part, we will be addressing each of them individually, focusing on the substantial progresses that microfluidics has contributed to produce within these research fields, in the last decades.

6.2.2.1 Haptotaxis

At the leading edge, migratory cells explore the surrounding microenvironment projecting actin-driven lamellipodia on both sides of the cell main axis. Within them, filopodia i.e., smaller cytoplasmic actin protrusions, extend far beyond lamellipodia boundaries. They are shown to be fundamental in substrate sensing and promoting and partially directing the cell motion. When cells advance, the lamellipodia actin filaments push against a large number of integrin-based adhesive complexes, linking the substratum to the cell surface. The focal adhesions are simultaneously pulled from the back by myosin-induced forces. The more cell-substrate junctions are formed, the stronger is the force that can be generated by lamellipodia protrusions moving the cell forward, whereas fewer ones provide weaker anchorage. Myosin pulling forces indeed, have tendency to reduce the cell-substrate contact surface and therefore the number of adhesion complexes. Both sides of the cell engage a sort of tug of war, whose winner sets the direction of the motion. Haptotaxis is the cell capability to move up a gradient of adhesion sites (Carter 1967). The activity of integrin-based adhesions complexes, however, is not limited to mechanosensing but it extends to mechanotransduction, as signaling pathways are triggered and/or modulated by the interplay among different integrin-associated protein families, such as Src kinases and Rho GTPases (Geiger et al. 2009, Harburger and Calderwood 2009). It is then problematic to discern which part of the cell motion is driven by haptotaxis, from that induced by the integrin-based receptors activation and the corresponding signaling cascade. The latter however, differs from chemotaxis by the fact that chemoattractants are expressed or bound on the surface rather than in a soluble fluid (Moreno-Arotzena et al. 2014).

The haptotaxis systematic investigation has received a considerable boost from microfluidics in recent years. Gradient microgenerators can produce with high accuracy degrading chemoattractant concentrations, that can be engraved in the substratum by protein adsorption (Jiang et al. 2005a, b). In an elementary T-junction microfluidic device indeed, two solutions, injected one against the other, combine at the junction forming a gradient of chemotactic bioagent at the perpendicular T arm, diffusing across the solutions-induced laminar flow. Then, the process of biomolecule adsorption onto the substratum is very rapid, owing to the large surface-to-volume ratio typical of microfabricated devices (Caelen et al. 2000). Devices features such as shape, geometry and cross-section, as well as the period of surface exposition to solutions, can be easily adjusted during the design and fabrication processes in order to achieve the desired gradients pattern of bound chemoattractant to the surface (Park et al. 2010a, b). This kind of technology, although simple and fast, displays shortcomings which have the undesired effects of compromising the designed symmetry of the imprinted pattern on the substrate. These can be air bubbles or other type of micro-hindrances disturbing the normal flow transit throughout the device, but also the possibility of surface saturation by chemoattractants, leading to the formation of proteins multilayers at the interface. A simple solution to this problem has been recently suggested in Hsu et al. 2005, were a new technique of collagen micropatterning could produce step changes of the adsorbed molecules density. Morevor, during the last decades, the blooming of new microfabrication techniques have allowed the production of cellular niches or, more generally three dimensional microstructures, faithfully conformal to cell environments (Yeo et al. 2011; Wu et al. 2013; Lee et al. 2012). Correspondingly, properly designed diffusants microgenerators, can shape the adsorbates gradient profile on the 3D walls.

6.2.2.2 Durotaxis

As outlined in the previous sections, a cell attached to the ECM, or in general to an adhesive substrate, generates internal actin- and myosin-driven contractile forces necessary for its motility. Cell migration indeed relies upon these actin-myosin active forces, applied to the ECM via focal adhesions: adhesion, contraction and a final burst-pushing-forward constitute indeed the three distinct phases of any cell motion. The same forces, however, produce an internal stress, called cytoskeletal pre-stress, responsible for the considerable stiffening of adherent cells (Wang et al. 2002; Kollmannsberger et al. 2011). In general, the internal pre-stress is the necessary requirement for the migratory dynamics activation but, nevertheless, it has the undesired drawback of cell stiffening, as they ultimately are ascribable to the same origin: it has been undoubtedly demonstrated the linear relationship among them, save for a systematic offset. A stiffer cell can hardly deform or move, displaying slower or no migration at all. This conflict recalls, *in nuce*, what ordinary skeletal muscles face while contracting: higher forces produce lower shortening velocities, while forces are almost absent during fast shrinking. This is not surprising if one considers the common acto-myosin nature of cells as well as muscle forces (Hill 1938). The rigidity of the substrate plays, in this respect, a pivotal role since stiffer substrates support stronger contacts at the cell adhesive interfaces, allowing the build-up of larger traction forces and a large spreading (Discher et al. 2005). A soft substrate on the other side, slipping under mechanical load, will be less efficient and will largely deform, preventing the cell to create the necessary conditions for the force accumulation and ensuing spreading. Stiffness-dependent spreading explains, at least from a purely mechanical point of view, why cells are more polarized on the stiffer substrate, and direct their migration from softer regions to those characterized by higher stiffness, while simultaneously decreasing their velocity, a process called durotaxis (Harland et al. 2011; Lange and Fabry 2013; Feng et al. 2018).

The contribution furnished by microfluidic to the understanding of the intimate connection existing between cell migratory dynamics and internal tensile self-generated forces, has been substantial. Common assays offer direct or indirect ways of inducing cell deformation, such as traction force microscopy on micropillars, microneedle posts and microcontact arrays (Tan et al. 2003), or on a compliant gel encapsulating fluorescent beads (Han et al. 2012). Besides characterizing the cell morphology by varying the density, stiffness and size of adhesion sites, these techniques can also quantify the magnitude of traction forces generated by cells. By instance, when micropillars are run over by a growing cell, they bend as a consequence of the traction forces applied on their tips and surrounding regions. Hence, they can be used as cantilever to quantify the magnitude of these forces, by simply monitoring their displacement. Two are the ingredients affecting such measurement: the microfabricated pillars geometry and the material stiffness. By varying each of them, one obtains devices within a wide degree of stiffness. The force estimate is relatively easy and it does not rely on any reference image, as the pillars reference positions are set up by the array unperturbed configuration, which is known. This type of studies have demonstrated on one hand that the spreading of the cell correlates to a certain amount of gene up- and down-regulation and, on the other, they allowed a precise characterization of the mechanical conditions contributing to the mechanotransduction response and biochemical pathways activation (Dalby et al. 2005). As stressed, strong adhesive attachments on a flat 2D stiff matrix have the effect to promoting the cell spreading and large traction forces, but they are also required for the cell endocytosis, shape stability and resistance against fluid shear stress. However they also make the cell stiffer, suppressing considerably its migratory propehension. The opposite happens in 3D microenvironments, like a dense 3D biopolymer network, through which cells migrate, by sensing a stronger hindrance to spreading and motion than in 2D systems. In this situation strong focal adhesions and high traction forces are needed to pull the cell within the network and to move the matrix fibers away, thus propelling rather than slowing down the motion.

As a matter of fact, recently it has been shown as for cells migrating through a 3D microfabricated biopolymer fiber matrix, direction, polarization and cell orientation are positively correlated with the magnitude of detected forces (Koch et al. 2012).

6.2.2.3 Plithotaxis

Cells move in the interstitial space of tissues and organs. Therefore, understanding how the micro-environment geometry and confinement influence their motion constitutes a central question. In this respect, the use of versatile microfluidic devices, with varying geometry, can elucidate some basic mechanism of cells motility and helps to extract the physical parameters controlling the migration. The systematic analysis of the experimental images from time-lapse movies reveals physical quantities like cell and boundaries velocity, conformational and directional dynamics, fluctuation of speed. By instance, varying the confinement degree, the question of the cell velocity in large and narrow interstitial environments is practically addressed: cells move faster when they are in 2D-confinement as compared with a non-confined 2D-environment, but migrate slower when they move through narrow constrictions. In this respect, when migration is restricted in one dimension (channels), fast and persistent motion of individual cells is observed for several hours in the absence of external chemical gradients (Irimia et al. 2009). Furthermore, the asymmetry of the patterning may be a discriminating factor: in a teardrop-shaped microdevice, a cell will protrude its lamellipodia from the sharp to the blunt end and, once released, it will likely move toward the blunt end direction (Jiang et al. 2005a, b). In artificial confinements, a cell will adopt the same motile strategies as migrating through a sheet of identical cells. In this case, traction forces deriving from cell-cell interaction must be added to the usual cell-ECM focal adhesions. The contribution given by microfluidics in this research domain has been substantial: the elaborated velocity field and the inferred force field can be evaluated at the intracellular contact points, with such an accuracy that normal and shear components can be enucleated. (Ladoux 2009; Trepat et al. 2009; Trepat and Fredberg 2011; Tambe et al. 2011). Forces between cells display an extremely high space-time heterogeneity, which is spread everywhere throughout the monolayer, i.e. it is not specific to some cell or group of cells. The stress arising from the single cell traction forces propagates to the adjacent cells via surface contacts, yielding to a long-ranged stress alignment that can span over multiple cell sizes. The motion of a single cell is due to the stress imbalance rather than to the experienced stress magnitude, thus explaining the observed dynamical heterogeneity. However, it happens that the largest stress is measured in cells where the imbalance is also the largest. As a consequence, the cell motion is driven by the stress anysotropy, a process called plithotaxis.

6.2.3 Electrotaxis

It is known that cells move in a direct current electric field, although not all migrate in the same direction, neither respond to the electrical stimulus in the same way (Robinson and Kenneth 1985; Swami et al. 2009). This type of migratory motion goes under the name of electrotaxis or galvanotaxis(Cortese et al. 2014). A remarkable example is furnished by the epithelial cells, whose electrotactic migration is stimulated by wound healing. An electric potential difference indeed is generated crosswise the epithelial layers, with the basal lamina functioning as anode and the upper part as chatode. This electric potential is called trans-epithelial potential (TEP). A wound damage causes disruption of the epithelial barrier, therefore creating a low resistive shortcut between the epithelium layers, whose high resistance in normal conditions would prevent any cell motion. The wound acts as a low-resistance cathode, thus attracting the epithelial cells which are positively charged. The sensing machinery involved in electromigration is still unclear, whereas there is partial evidence that ion channels and surface receptors dynamics could take active part in it (McCaig et al. 2005). In spite of this, electrotaxis is heavily influenced by the surrounding conditions such as medium concentration, and it does not occur in any type of cell nor invest the activation of similar signaling pathways.

In recent years microfluidics has provided the suitable platform for the efficient investigation of the cell electrotactic activity (Li and Francis 2011). In these assays, an uniform electric field is applied to cells confined in microchannels, monitoring their migratory activity by real-time imaging. Moreover, in microfabricated channels the Joule heating is drastically suppressed due to the devices dimensions, allowing high-throughput experiments to be carried out easily. This is a considerable advancement, if one considers the shortcomings arising in the conventional Petri dishes, including the absence of miniaturization and the lack of spatial control of the electric field. Finally, as the manufacturing process is analogous to chemotactic assays, chemotaxis and electrotaxis migration stimuli can be superimposed and the their interplay studied and quantified on the same chip, with a particular focus on the diversity of effects that they can produce on cell migration (Li et al. 2011; Wang et al. 2011).

6.3 Invadosome Formation in Confined Environments

Invadosomes are actin-rich adhesive structures that mediate matrix degradation, mechanosensing, cell migration and invasion (Genot and Gligorijevic 2014). Therefore, their role in normal development as well as in disease progression is of fundamental importance (Paterson and Courtneidge 2018).

Invadosomes are formations common to many cells, such as macrophages, osteoclasts, vascular smooth muscle cells, endothelial cells, Src-transformed fibroblasts, and cancer cells (See Fig. 6.2a, b). More commonly, by definition, invadosomes refer to a group of punctate structures that can be found in normal cells (known as podosomes) and in cancer cells (invadopodia). Their diameter ranges between 0.5 μm and 2 μm, whereas the length is 0.2 μm–0.4 μm in the case of podosomes, and 2 μm–5 μm for invadopodia (Linder 2007). Invadosomes consist of a core structure where a dense filamentous (F)-actin is associated with actin regulatory proteins such as Arp2/3, WASP/N-WASP, cortactin. The actin-rich core is surrounded by adhesive clusters which contain integrins, vinculin, talin and other focal adhesion proteins (Linder 2007) (Fig. 6.2c). The scaffolding and Src substrate Tks5 (SH3PXD2A) is a marker of the structures. In addition, podosomes in macrophages and dendritic cells contain a substructure, called cap structure, that is proposed to serve as a hub for incoming vesicles and might be involved in podosome-associated contractility (Bhuwania et al. 2012; Linder and Wiesner 2015; Cervero et al. 2018).

Podosomes form an interconnected network that allows a collective behavior and communication between single protrusions (Luxenburg et al. 2007; van den Dries et al. 2013). At the same time, connections between individual podosomes entail the efficient organization into higher-order structures like clusters, rosettes, rings or belts (Veillat et al. 2015). This organization pattern is absent for invadopodia (Fig. 6.2a, b). They are also fewer in number, approximately between 1 and 10 per cell, in striking contrast with podosomes that range between 10 and several hundred (Linder 2007).

Invadosomes differ from other adhesive cell structures, such as focal or fibrillar adhesions by their proteolytic capacities mediated by metalloproteases (Linder 2007). This inherent ability to degrade the ECM allows cells in restricted areas to create room for new biological structures, opening paths for cells to migrate. Podosomes, in general, enable a broader and shallower degradation of the surrounding matrix, while invadopodia induce a more focused and deeper penetrating ECM degradation (Fig. 6.2c). This is crucial in migrating processes like cancer cell extravasation, leukocyte trafficking, and angiogenic sprouting, namely for those biological processes characterized by the crossing of anatomical boundaries (Genot and Gligorijevic 2014). For instance, the endothelial podosomes appear to be the critical regulators of developmental sprouting angiogenesis *in vivo* through breaking vascular basement membrane (BM) (Spuul et al. 2016b). In addition, endothelial podosome rosettes have been shown to control the branch-

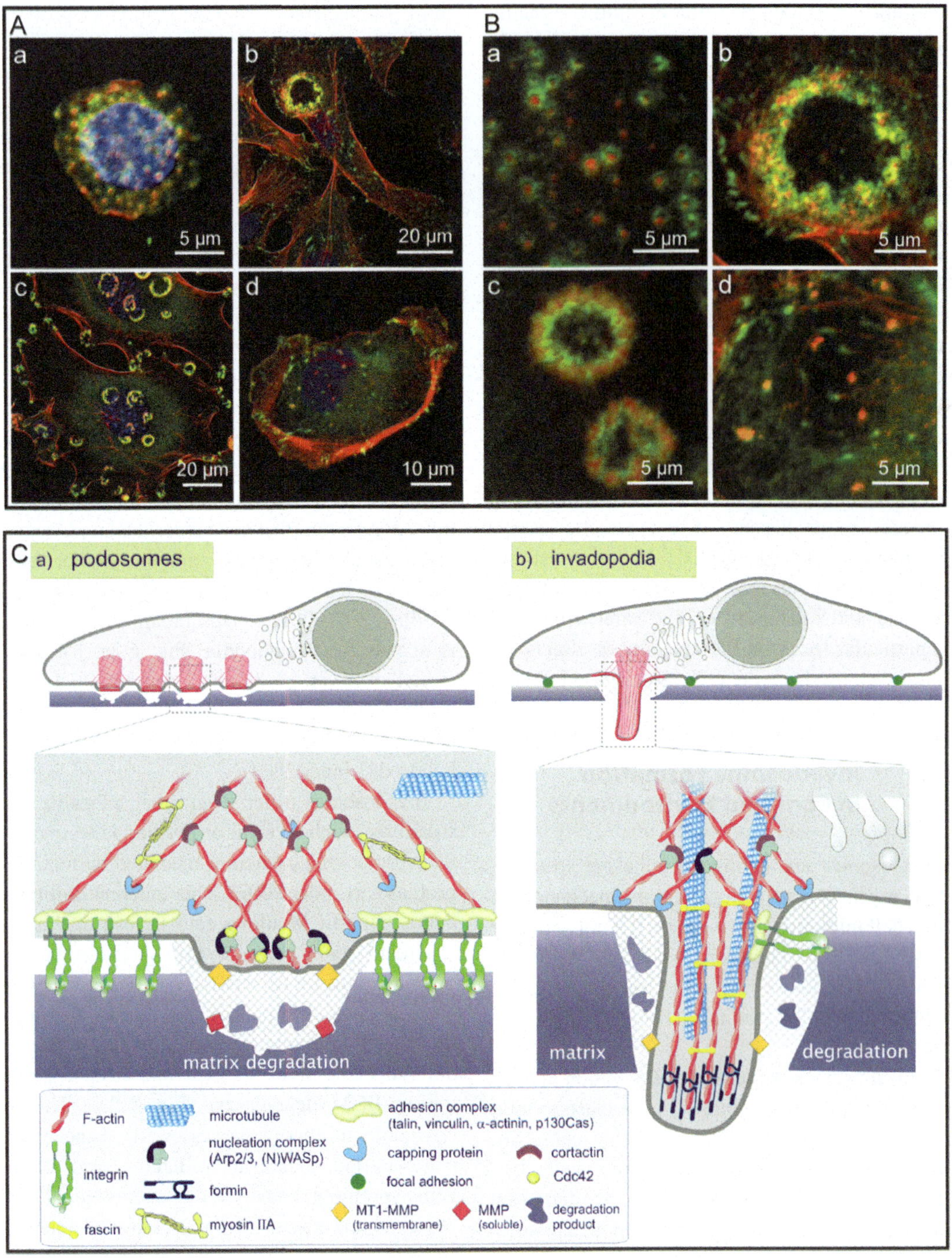

Fig. 6.2 Invadosome architecture, patterning and schematic representation. (**a**) Invadosomes form in different cell types and show great diversity in subcellular arrangements. (*a*) A stationary macrophage with individual podosomes covering the majority of the substrate-attached cell side; (*b*) a TGFβ-stimulated endothelial cell with an interconnected podosomes forming a rosette superstructure; (*c*) Rous sarcoma virus (RSV)-transformed fibroblasts with numerous invadosome rosettes; (*d*) a cancer cell with invadopodia located in the vicinity of the nucleus. (**b**) Invadosome architecture and patterning in cells corresponding to panel A shown at higher magnification. (**a**, **b**) Invadosomes are visualized with F-actin (red) and vinculin (green) staining and nucleus is highlighted in blue. (**c**) Schematic representation of invadosomes: podosomes (*a*) and

ing of blood vessels in a model of pathological angiogenesis (Seano and Primo 2015).

Invadopodia characterize the migration phase of cancer cells, whereas podosomes appear spontaneously in myelomonocytic cells or upon stimulation in non-hematopoietic cells, like endothelial cells (ECs) (Veillat et al. 2015). Generally speaking, invadosomes can be induced by different growth factors like TGFβ (Varon et al. 2006), VEGF-A (Daubon et al. 2016); matrix components Collagen (Juin et al. 2012), Fibronectin (Spuul et al. 2016a), Collagen-IV (Daubon et al. 2016), mechanical cues (Spuul et al. 2016a), bacterial infection (Le Roux Goglin et al. 2012), and a variety of other signaling pathways and microenvironmental cues.

Because of their active part in cell migration, the past 3 years have witnessed a considerable thrive of research on podosomes and invadopodia, including the discovery of novel components, the characterization of stimuli required for their formation, and the establishment of their *in vivo* relevance. Formation of invadosomes under confinement instead, is a research topic that has come up only recently. The understanding of how the microenvironmental and the surrounding mechanical cues promote (or suppress) invadosome formation and the ensuing alterations in their dynamics, constitutes the powerful fly-wheel for the research activity in this field. Experimental studies in this new-born area of investigation are still limited, and include *in vivo* studies or *ex vivo* studies based on the use of transwell devices or involving 2D, quasi-2D confinements and microfluidics experiments (see Fig. 6.3 for a schematic representation of the different confinement environments).

In the next sections we will be reviewing in details the aforementioned studies, going, when possible, through podosomes and invadopodia separately.

6.3.1 Invadosomes Formation in *in vivo* Environments

Cells observed *in vivo* are surrounded by the native ECM and bio-molecules, and naturally restricted from the best fit environment. For instance, endothelial cells (ECs), which line the inner surface of all blood vessels, are exposed to many mechanical cues like confinement-induced constrictions, shear stress, ECM composition and rigidity, vascular topography *in vivo*: they respond to these cues by adapting their behavior. However, the *in vivo* setup comes with several drawbacks including practical difficulties of comparing the response of target cells exposed to different drugs or reagents in parallel, and to observe the invadosomes dynamics in real time. As a result, the general protocol is to perform immunohistochemical examination of the samples, so that *ex vivo* and truly *in vivo* studies are often used as synonyms. As a first example, two invadopodia-related proteins, namely actinin-1 and cortactin, are found colocalized at the matrix–contact-side in lung adenocarcinoma cells, and these observations were extrapolated to the metastatic invasion in patients which did not undergo a preoperative chemotherapy (Hirooka et al. 2011).

Confocal microscopy is the privileged tool for the observation of samples with limited thickness, around 250 μm, depending on the lens magnification and sample transparency. Mouse aortic explants exposed to TGFβ *in vitro* were observed by confocal microscope, showing that the endothelial podosome rosettes

Fig. 6.2 (continued) invadopodia (*b*). (*a*) Podosomes are columnar structures that extend upwards from ventral cell surface into the cytoplasm and have a core of F-actin associated with Arp2/3/WASP/Cdc42 driving the actin polymerization. The core is surrounded by and linked to the adhesive clusters containing integrins and other focal adhesion proteins through radial actin filaments. (*b*) Invadopodia penetrate into the ECM as long filopodial-like membrane extensions having the Arp2/3/N-WASP complex at the base of the invadopodia. The adhesive clusters do not form a ring around the core in invadopodia. (Figure reproduced from (Spuul et al. 2014) with permission from Taylor & Francis Ltd. (www.tandonline.com))

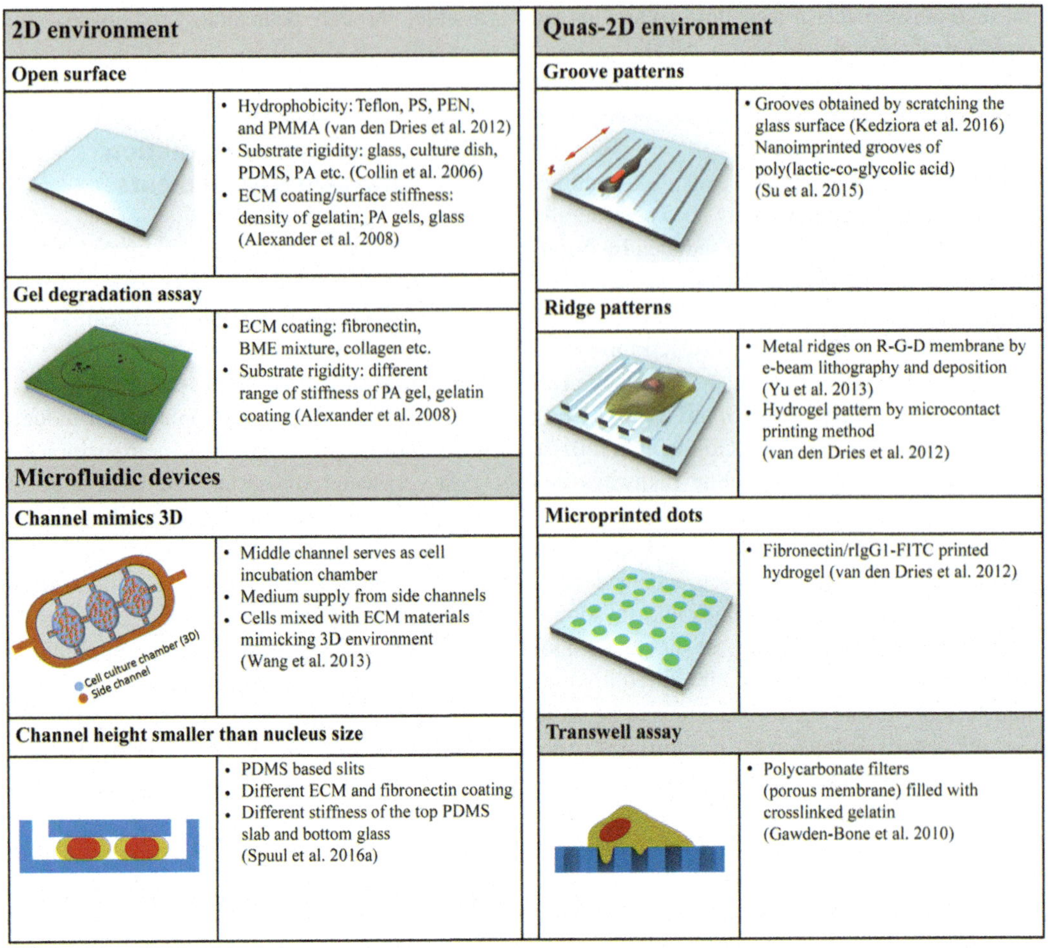

Fig. 6.3 Experimental setups for the study of invadosomes formation and dynamics. *PS* polystyrene, *PEN* polyethylene naphthalate, *PMMA* polymethyl methacrylate, *PDMS* polydimethylsiloxane, *PA* polyamide, *ECM* extracellular matrix, *BME* basal medium eagle, *FN* fibronectin, *R-G-D* tripeptideArg-Gly-Asp

observed under TGFβ stimulation in ECs in the culture dish, could be also visualized in the endothelium of native arterial vessel. Moreover, basement membrane Collagen-IV degradation was observed under the podosome rosettes (Rottiers et al. 2009). Similar results were obtained with mouse aortic explants exposed to angiogenic stimulation VEGF where podosome rosettes were detected in the vascular angiogenic endothelium on the vessel side (Seano and Primo 2015).

Demonstration of podosomes in the native tissue in situ has been described during angiogenesis in the retina mouse model (Spuul et al. 2016b). VEGF-A/Notch-regulated podosomes in angiogenic ECs were shown to degrade the basement membrane Collagen-IV facilitating sprouting within the developing vasculature. Importantly, the podosomes seen in microvascular tip cells present as single F-actin/Cortactin/P-Src rich globular structures, thus morphologically distinct from cultured EC 2D podosomes arranged in rosettes. However, more distally, the podosomes detected in angiogenic large vessels appeared as interconnected single dots or sometimes in clusters. Therefore, the study highlighted the importance of the microenvironment in regulating the arrangement of podosomes in angiogenic ECs (see also below Sect. 6.3.4).

Moreover, intravital microscopy, which enables observation of dynamic biological processes, and a depth in the range of tens of microns to less than 2 mm underneath the biological sample surface, has allowed *in vivo* imaging of invadosomes activities (Andresen et al. 2012; Masedunskas et al. 2012; Gabriel et al. 2018).

In vivo, the lesions occurring in cardiovascular diseases or tumors are characterized by fibrosis. Interestingly, these fibrotic lesions, associated with stiffened ECM (Kai et al. 2016), promote aberrant cellular mechanotransduction and enable resident cells migration by the assembly of the invadosomes and lamellae.

6.3.2 Transwell Assays

Biological barriers are omnipresent in *in vivo* environments. It is however difficult to set well-controlled physical or chemical environments reproducing the heterogeneity found in natural systems. Thus, engineering microenvironments for the analysis of cells migrating in a constricted space, and using a variety of biomaterials, has become a fundamental issue. This has led to study the intimate relationship between the invadosome machinery and cell migration across physical barriers, providing a better understanding of cell behavior in real physiological or pathological situations. Transwell devices are commonly used to study cell migration and invasion (Kramer et al. 2013), as membrane with various pore sizes are commercially available. The transwell method has been also applied to the invadosome dynamics studies, where porous membranes can be coated with ECM proteins (Fig. 6.3). In one study, immature dendritic cells (iDCs) were seeded on polycarbonate filters that had been coated with crosslinked gelatin (Gawden-Bone et al. 2010). Invadosomes in iDCs were shown to detect the pores and adopt a protrusive architecture where the actin core engaged into the pore. The matrix degrading activity was shown to depend on the matrix metalloproteinase MMP-14 by using both electron microscopy and three-dimensional structured illumination microscopy.

6.3.3 Invadosomes Formation in 2D Environments

Many factors, such as the stiffness of the ECM, the rigidity of the substratum, the surface materials and geometry, and other mechanical cues, influence invadosomes formation and dynamics (Fig. 6.3). The cellular response to these environmental cues has been mainly studied in strictly 2D contexts that ease the experimental preparation and invadosomes detection. It should be kept in mind that podosomes morphology assessed in 2D setups markedly differs from that in 3D microenvironments and from that visualized *in vivo*. Likewise, invadosome dynamics is expected to be significantly different in 2D and 3D contexts.

Different substrate materials such as glass cover slides or culture dishes can be used, coated or not with ECM proteins. The invadosomes response to hydrophobicity has been addressed by varying the substrate material: Teflon, PS, PEN and PMMA were used in the 2D environments, ranging from the most hydrophobic (Teflon) to the most hydrophilic one (PMMA) (van den Dries et al. 2012). Remarkably, in these experiments, the formation of invadosomes in iDCs was not significantly affected by substrate hydrophobicity.

Mechanical cues alter both cell migration mode and cell morphology in 2D, and similarly invadosomes activity. In one study (Lo et al. 2000), the term durotaxis was coined: 3 T3 cells on collagen-coated polyacrylamide sheets, where a gradient of rigidity was produced, exhibited a preferential motion toward the stiffer substrate. Six years later, this tendency was shown to be connected with the activity of invadosomes. On similar collagen-I-coated polyacrylamide substrates with varying stiffness, another study explored for the first time the dynamics of invadosomes formed by GFP-actin transfected NIH-3 T3 fibroblasts (Collin et al. 2006). The transition in rigidity on the polyacrylamide substrate was introduced by controlling the ratio between monomer and cross-linker concentrations (Pelham and Wang 1998). Two different ratios of bisacrylamide/acrylamide

(0.2%, 0.08%) were used and invadosome dynamics on these substrates was compared with the corresponding situation on glass (defined as the reference rigid substrate). It was found that increasing substrate stiffness leads to a remarkable increase in podosome lifespan. Experiments have also pointed out how the podosome rosettes appear more blurred on soft substrates than on rigid ones. At the same time, increasing substrate rigidity, decreased the mean distance between neighboring podosomes within the rosette (Collin et al. 2006).

The interplay between ECM density and substrate rigidity also affects invadosomes functionality, as demonstrated by ECM degradation assays (Alexander et al. 2008). In this type of assays, coating is typically achieved by fluorescence dye-ECM materials: that allows quantitative analysis as the surface of the degraded matrix can be assessed (i.e loss of fluorescence). CA1d breast carcinoma cells were cultured on ECM substrate of 0.5–5% gelatin with storage modulus ranging from 61 to 1190 Pa, coated with FITC-labeled fibronectin (FITC-FN). Invadopodia increased in numbers and the ECM-degraded area was increased on denser gelatin substrates (2.5–5% gelatin), showing that the rigidity promoted invadopodia proteolytic activity. However, increasing the gelatin concentration could also increase the number of integrin binding sites and integrin ligation which is known to promote invadopodia formation. CA1d breast carcinoma cells cultured on stiffer (3300 Pa) polyacrylamide gels, with 1% of gelatin and FITC-FN coating, produced even more invadopodia with increased ECM degradation capabilities, as compared to cells cultured on softer gels (360 Pa) (Alexander et al. 2008).

Another way of confining the invadosomes in a quasi-2D microenvironments, and thereby study how the geometry influences their formation, is to force them into patterned surfaces (Fig. 6.3). The topological pattern can be manufactured by elementary engineering methods, such as the scratching of the glass surface in order to produce straight grooves. In (Kedziora et al. 2016), 375 M (melanoma) cells expressing active c-Src (SrcY530F) were seeded on scratched glass coverslips, and the upforming invadosomes were found to protrude along the patterned grooves. The same alignment dynamics in length and direction of retraction fibers was observed during fibroblast cell division (Su et al. 2015). Another method to create 1D patterns in a 2D system is the deposition of metal lines on the tripeptideArg-Gly-Asp (RGD) membranes. In the experiment reported in ref. (Yu and Groves 2010), 100 nm width metal ridges were fabricated by e-beam lithography and thin film deposition. Since lipid molecules can only self-assemble on the glass substrate, the ridges could then serve as metal borders intercalated between RGD stripes. Fibroblasts were seeded on top of the RGD membranes separated by the metal ridge patterns, and the ensuing invadosomes formation was observed to depend critically on the ridges density and their relative spacing. Cells formed less percentage of invadosomes when spreading on denser patterning on RGD membranes (Yu et al. 2013), suggesting that line spacing and density influence trivially on the invadosomes activity. When DC were seeded on micropatterned substrates, with grooves of various heights and width ranging from 2 to 20 μm, the podosomes were observed to invariably align along the edges notwithstanding the shape of the pattern or of the physico-chemical substrate characteristics (van den Dries et al. 2012). Moreover, podosomes were predominantly formed on the edge of the patterns.

The microcontact printing method is another frequently used method to generate multiple quasi-2D patterns by PDMS stamp (van den Dries et al. 2012). By this method, circular patterns of fibronectin/rIgG1-FITC printed hydrogel were generated, and iDC were seeded on spots of diameters in the range 5 μm–20 μm, with a distance between spots between 7.5 and 10 μm. Interestingly, it was shown that the number of invadosomes formed on the spots directly correlated with their size. Most intriguingly, the number of invadosomes spreading on different spot size remained quite stable.

6.3.4 Invadosomes Formation in Microfluidic Devices

The need for well-controlled experimental conditions, e.g., equal exposure of cells to drugs and reagents, or reproducible geometric environments and chemical conditions, strongly promoted the use of microfluidic devices to unveil the invadosomes functionalities in confined space (Fig. 6.3).

The study reported by Li and colleagues showed how to quantify the migration of macrophage-assisted cancer cells inside a device consisting in 3 parallel microchannels (Li et al. 2017). The middle channel was loaded with macrophages and cancer cells together with collagen-I gel, with both ends bathing in the culture medium. The posts lining between the boundaries of the parallel channels, not only prevented the channel from collapse, but also ensured diffusion of nutrients from the culture medium. Cells were thereafter monitored while migrating into the channel. This type of microfluidic device aims at mimicking a 3D micro environment, where the incubation channel is 120 μm in height and 1300 um in width, and completely filled with ECM (Fig. 6.4a). Similarly, although with a slightly different design, microfluidics channels also promote invadopodia formation in lung cancer cells, A549 cells (Wang et al. 2013). In this case, the middle microchannel served as cell culture chamber for cancer cells-BME mixture, and connected to the two lateral channels for drug application or medium addition. These microfluidic devices could actually mimic the 3D microenvironment while reagent can still be applied in a tunable fashion (Fig. 6.4b). Thanks to this design, parallel tests of invadosomes formation in EGF and GM6001 (MMP inhibitor)/EGF cells, against a control group of cells, could be carried out synchronously. The EGF groups exhibited the highest percentage of cells forming invadosomes. Moreover, it was shown that the morphology of

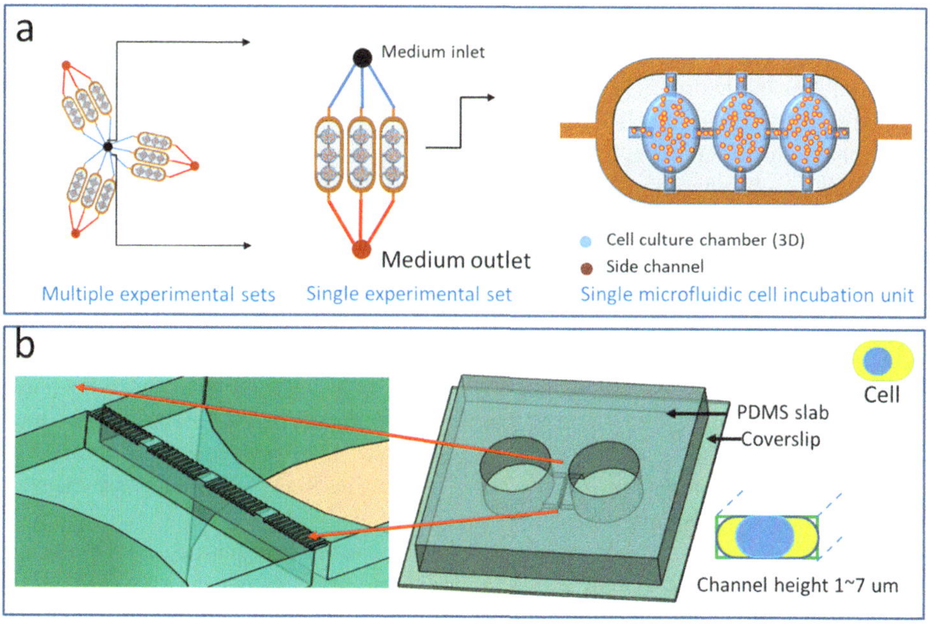

Fig. 6.4 Microfluidic devices for the study of the invadosomes dynamics. (**a**) Microfluidic design for the device used in ref. (Li et al. 2017) to investigate cancer cells invasion in 3D matrix and simultaneous invadosome formation. The middle cell incubation chamber serves like 3D environment for the cells. Side channels can provide the nutrition, reagents, or drugs for the middle incubation chamber. This simple design allows multiple experimental tests. (**b**) Simple microfluidic design for the experiment in (Spuul et al. 2016a). Invadosomes formation and dynamics was observed in restricted environment characterized by a channel height smaller than the nucleus size

A549 cells is drastically affected by the confining environment. Indeed cells cultured on 2D culture dishes, were compared to cells grown within the 3D ECM matrix inside the middle channel. The A549 cells in 3D matrix were round in shape, while those grown in 2D environments attained a more elongated conformation.

In the process of metastasis, tumor cell migration through confined spaces is facilitated by MT1-MMP-mediated proteolysis of confining fibrils (Castro-Castro et al. 2016). Dissolution of the matrix is required for nuclear translocation as nuclear stiffness has been shown to be the limiting factor during confined migration (Wolf et al. 2013). Using the model of MDA-MB-231, human breast cancer cell line, Infante and colleagues showed that, upon intact nucleo-cortex linkage, confined migration triggered the polarization of MTI-MMP rich endosomes in front of the nucleus (centered by centrosome) (Infante et al. 2018). Here MTI-MMP rich invadopodia assembled and mediated the degradation of the surrounding collagen fibrils. The authors called this response as ´digest on demand´ as the degradative invadopodia were assembled upon demand (mechanical signal) and used to digest the matrix. This pericellular collagenolytic activity promoted the nuclear movement in confined microenvironment and supported tumor cell migration. To create a 3D confining environment, tumor cells were embedded in a fibrillar collagen-I network with varying pore sizes. The authors proposed the ´digest on demand´ response through the assembly of MT1-MMP-rich invadopodia as a novel mechanism for tumor cells to migrate through constricted 3D microenvironments (Infante et al. 2018).

EC migration is particularly important in the angiogenesis process, where the formation of new blood vessels from an existing vasculature occurs. In this context, the nucleus deformability is an important factor because cells are penetrating gaps much smaller than the whole cell size. The nucleus of a cell is known to be stiffer than the cytoplasm and, therefore, it is of fundamental importance to understand how cells react to confinement smaller than the nucleus size. This question was addressed for the first time in (Spuul et al. 2016a), where it was also shown that the use of microfluidic devices could offer the practical advantage of studying systematically and concomitantly EC migration and the rearrangement of their actin cytoskeleton induced by confinement, combined with the contribution of different ECM proteins (Spuul et al. 2016a).

Aortic ECs do not assemble podosomes spontaneously; however stimulation with TGFβ induces the formation of podosome superstructures called rosettes (Varon et al. 2006). Unstimulated bovine aortic ECs (BAECs) were subjected to varying degree of confinement in a microfluidic device. The PDMS-based chip used in the study contained parallel transversal channels (slits) ranging from 1.38 μm to 7.3 μm in height, 50 μm in width and 250 μm in length. Two open-top incubation chambers at each ends of the device, allowing cell seeding and reagent exchange, were connected with the channels through an intermediate channel (Spuul et al. 2016a). Surprisingly, BAE cells, although unstimulated, acquired the ability to form podosomes in the confining channels, a process that was shown to be stimulated by the fibronectin (FN) coating the device. FN indeed, enhanced the cell migration in confined conditions on the one hand, and guaranteed the cell viability on the other hand. Importantly, the podosomes assembled under confinement did not form rosettes but appeared instead in clusters or as individual dots, forming mostly on the cell forefront at the matrix contacting cell side. This suggests that confinement-induced podosomes might be used as devices to degrade the confining matrix to make room for the migrating cell. At the same time, it was observed that the percentage of BAECs forming invadosomes increased as the channel height decreased.

An intriguing dependence on the substrate stiffness was also detected: invadosomes formed more abundantly on the PDSM soft side (E = 580 kPa), than on the stiffer glass side (E = 7.29 × 10E7 kPa), when BAECs were migrating in shallower channel (1.28 um height). This is in contrast with the previous finding that invadosomes formation traces the durotactic migration tendency of the cells (Collin et al. 2006). This issue, however, needs further investigation since a straightforward comparison between the experiments reported in (Collin et al. 2006) and in (Spuul et al. 2016a) might be misleading, considering the differences in cell types, coating proteins and stiffness range.

In 2D environments, human microvascular ECs (HMVECs) derived from capillaries assemble podosome rosettes spontaneously in a small fraction of the cells (<6%). Moreover, after VEGF-A stimulation, an increase of both the matrix degrading potential of the assembled podosomes, and the number of cells forming podosomes (up to 15%) was observed. HMVECs in 3D confined conditions, and in the absence of VEGF-A, migrated into FN coated channels (1.38 μm–7.3 μm in height), they formed a pseudo-endothelium and assembled podosomes that were found scattered but interconnected by actin cables. Podosomes appeared at the plasma membrane-matrix contact sites and contained MT1-MMP, suggestive of a proteolytic activity (Spuul et al. 2016b). Interconnected podosomes detected in the microfluidic device resembled strongly the ones seen during *in vivo* vascular remodeling of large angiogenic vessels (Spuul et al. 2016b). These findings highlight the importance of microfluidic devices offering a possibility to study many microenvironmental cues in controlled environment, that reproduce mechanical and biochemical aspects of in vivo angiogenesis.

Cell migration in confinement has been studied also in human mesenchymal stem cells (hMSCs) (Doolin and Stroka 2018). Doolin and Stroka have shown that the cellular cytoskeleton (actin and microtubules) reorganizes its structure when cells are moving in microenvironments. To account for this finding, the authors have proposed a differential role for the cytoskeletal and contractile machinery in cell migration, while migrating in confined vs unconfined environments. Migration of hMSCs was studied in Collagen-I-coated microchannels with varying width (3–50 μm). In this case it was shown that the number of mature focal adhesions decreases by increasing the confinement.

Migration in confinement induces the assembly of mechanosensory invadosomes that transduce the mechanical stress applied to the cells into biochemical response. All together, these studies have established the fundamental role played by the confinement and ECM microenvironment in regulating the assembly of invadosomes, their architecture and spatial arrangement, and their capability of mediating matrix degradation, thus impacting cell migration and invasion.

Acknowledgements This work was supported in part by the AS Thematic Projects [AS-106-TP-A03] and the Ministry of Science and Technology (ROC) [105-2112-M-001-021-MY3, 106-2627-M-001-001, and 106-2119-M-001-005]. A.T. acknowledges the European Research Council through the Advanced Grant No.291002 SIZEFFECTS. P.S. acknowledges the support of TalTech Young Investigator grant B61, Estonian Research Council Starting Grant PUT1130 and G.F.Parrot Travel Grant. E.G. laboratory (http://genot-lab.org/) is funded by INSERM, the Ligue contre le Cancer, committee of the Gironde, the University of Bordeaux (transversal project HYPOXCELL) and the "Marfans" Association. The Invadosome Consortium is an international network of laboratories interested in adhesion structures involved in invasive processes. It is open to the entire scientific community (http://www.invadosomes.org/). We thank Chia-Tzi Kuo for the help in drawing Fig. 6.3. Confocal images of Fig. 6.1 were performed in part through the use of the advanced optical microscopes at Division of Instrument Service of Academia Sinica and with the assistance of Shu-Chen Shen.

References

Alberts B, Johnson A, Lewis J et al (2002) The extracellular matrix of animals. In: Molecular biology of the cell, 4th edn. Garland Science, New York

Alenghat FJ, Ingber DE (2002) Mechanotransduction: all signals point to cytoskeleton, matrix, and integrins. Sci STKE 119:pe6–pe6

Alexander NR, Branch KM, Parekh A et al (2008) Extracellular matrix rigidity promotes invadopodia activity. Curr Biol 18(17):1295–1299

Alt W, Deutsch A, Dunn G (2012) Dynamics of cell and tissue motion. Birkhäuser, Basel

Andresen V, Pollok K, Rinnenthal JL (2012) High-resolution intravital microscopy. PLoS One 7(12):e50915

Berthier E, Young EWK, Beebe D (2012) Engineers are from PDMS-land, biologists are from polystyrenia. Lab Chip 12(7):1224–1237

Bhuwania R, Cornfine S, Fang Z et al (2012) Supervillin couples myosin-dependent contractility to podosomes

and enables their turnover. J Cell Sci 125(Pt9):2300–2314
Brassard D, Clime L, Kebin L et al (2011) 3D thermoplastic elastomer microfluidic devices for biological probe immobilization. Lab Chip 11(23):4099–4107
Budday S, Nay R, de Rooij R et al (2015) Mechanical properties of gray and white matter brain tissue by indentation. J Mech Behav Biomed Mater 46:318–330
Caelen I, Bernard A, Juncker D et al (2000) Formation of gradients of proteins on surfaces with microfluidic networks. Langmuir 16(24):9125–9130
Castro-Castro A, Marchesin V, Monteiro P, Lodillinsky C, Rosse C, Chavrier P (2016) Cellular and molecular mechanisms of MT1-MMP-deis now present in the text.Pendent cancer cell invasion. Annu Rev Cell Dev Biol 32:555–576
Carter SB (1967) Haptotaxis and the mechanism of cell motility. Nature 213(5073):256–260
Cervero P, Wiesner C, Boussou A et al (2018) Lymphocyte-specific protein 1 regulates mechanosensory oscillation of podosomes and actin isoform-based actomyosin symmetry breaking. Nat Commun 9(1):515
Chang HC, Yeo LY (2009) Electrokinetically driven microfluidics and nanofluidics. Cambridge University Press, Cambridge, NY, pp 379–406
Charati SG, Stern SA (1998) Diffusion of gases in silicone polymers: molecular dynamics simulations. Macromolecules 31(16):5529–5535
Chaurey V, Block F, Su YH et al (2012) Nanofiber size-dependent sensitivity of fibroblast directionality to the methodology for scaffold alignment. Acta Biomater 8:3982–3990
Chiu DT, deMello AJ, Di Carlo D et al (2017) Small but perfectly formed? Successes, challenges, and opportunities for microfluidics in the chemical and biological sciences. Chem 2(2):201–223
Collin O, Tracqui P, Stephanou A et al (2006) Spatiotemporal dynamics of actin-rich adhesion microdomains: influence of substrate flexibility. J Cell Sci 119(Pt 9):1914–1925
Cortese B, Palama IE, D'Amone S et al (2014) Influence of electrotaxis on cell behaviour. Integr Biol 6(9):817–830
Dalby MJ, Riehle MO, Sutherland DS et al (2005) Morphological and microarray analysis of human fibroblasts cultured on nanocolumns produced by colloidal lithography. Eur Cell Mater 9(1):1–8
Daubon T, Spuul P, Alonso F et al (2016) E VEGF-A stimulates podosome-mediated collagen-IV proteolysis in microvascular endothelial cells. J Cell Sci 129(13):2586–2598
Discher DE, Janmey P, Wang YL (2005) Tissue cells feel and respond to the stiffness of their substrate. Science 310(5751):1139–1143
Doolin MT, Stroka KM (2018) Physical confinement alters cytoskeletal contributions towards human mesenchymal stem cell migration. Cytoskeleton 75(3):103–117
Duffy DM, Garrett SM, Ellis SE et al (2008) Influence of supramammary lymph node extract on in vitro cell proliferation. Cell Prolif 41(2):299–309
Englert DL, Manson MD, Jayaraman A (2009) Flow-based microfluidic device for quantifying bacterial chemotaxis in stable, competing gradients. Appl Environ Microbiol 75(13):4557–4564
Evstrapov AA (2017) Micro-and nanofluidic systems in devices for biological, medical and environmental research. J Phys Conf Ser 917(2). IOP Publishing
Feng J, Levine H, Mao X et al (2018) Stiffness sensing and cell motility: durotaxis and contact guidance. bioRxiv:320705
Gabriel EM, Fisher DT, Evans S et al (2018) Intravital microscopy in the study of the tumor microenvironment: from bench to human application. Oncotarget 9(28):20165
Gawden-Bone C, Zhou Z, King E et al (2010) Dendritic cell podosomes are protrusive and invade the extracellular matrix using metalloproteinase MMP-14. J Cell Sci 123(pt9):1427–1437
Geiger B, Spatz JP, Bershadsky AD (2009) Environmental sensing through focal adhesions. Nat Rev Mol Cell Biol 10(1):21–33
Genot E, Gligorijevic B (2014) Invadosomes in their natural habitat. Eur J Cell Biol 93(0):367–379
Gravesen P, Branebjerg J, Jensen OS (1993) Microfluidics-a review. J Micromech Microeng 3(4):168
Gu J, Gupta R, Chou CF, Wei Q, Zenhausern F (2007) A simple polysilsesquioxane sealing of nanofluidic channels below 10 nm at room temperature. Lab Chip 7(9):1198–1201
Hamm HE (1998) The many faces of G protein signaling. J Biol Chem 273(2):669–672
Han SJ, Bielawski KS, Ting LH et al (2012) Decoupling substrate stiffness, spread area, and micropost density: a close spatial relationship between traction forces and focal adhesions. Biophys J 103(4):640–648
Handorf AM, Zhou Y, Halanski MA et al (2015) Tissue stiffness dictates development, homeostasis, and disease progression. Organogenesis 11(1):1–15
Harburger DS, Calderwood DA (2009) Integrin signalling at a glance. J Cell Sci 122(2):159–163
Harland B, Walcott S, Sun XS (2011) Adhesion dynamics and durotaxis in migrating cells. Phys Biol 8(1):015011
Hartman CD, Isenberg BC, Chua SG et al (2017) Extracellular matrix type modulates cell migration on mechanical gradients. Exp Cell Res 359(2):361–366
Hepler JR, Gilman AG (1992) G proteins. Trends Biochem Sci 17(10):383–387
Hill AV (1938) The heat of shortening and the dynamic constants of muscle. Proc R Soc Lond B 126(843):136–195
Hirooka S, Akashi T, Ando N et al (2011) Localization of the invadopodia-related proteins actinin-1 and cortactin to matrix-contact-side cytoplasm of cancer cells in surgically resected lung adenocarcinomas. Pathobiology 78(1):10–23

Hsu S, Thakar R, Liepmann D et al (2005) Effects of shear stress on endothelial cell haptotaxis on micropatterned surfaces. Biochem Biophys Res Commun 337(1):401–409

Infante E, Castagnino A, Ferrari R et al (2018) LINC complex-Lis1 interplay controls MT1-MMP matrix digest-on-demand response for confined tumor cell migration. Nat Commun 9(1):2443

Irimia D (2010) Microfluidic technologies for temporal perturbations of chemotaxis. Annu Rev Biomed Eng 12:259–284

Irimia D, Geba DA, Toner M (2006) Universal microfluidic gradient generator. Anal Chem 78(10):3472–3477

Irimia D, Geba DA, Toner M (2009) Spontaneous migration of cancer cells under conditions of mechanical confinement. Integr Biol 1(8–9):506–512

Jiang X, Bruzewicz DA, Wong AP et al (2005a) Directing cell migration with asymmetric micropatterns. Proc Natl Acad Sci 102(4):975–978

Jiang X, Xu Q, Dertinger SK et al (2005b) A general method for patterning gradients of biomolecules on surfaces using microfluidic networks. Anal Chem 77(8):2338–2347

Juin A, Billottet C, Moreau VC et al (2012) Physiological type I collagen organization induces the formation of a novel class of linear invadosomes. Mol Biol Cell 23(2):297–309

Kai FB, Laklai H, Weaver VM (2016) Force matters: biomechanical regulation of cell invasion and migration in disease. Trends Cell Biol 26(7):486–497

Kedziora KM, Isogai T, Jalink K et al (2016) Invadosomes–shaping actin networks to follow mechanical cues. Front Biosci (Landmark Ed) 21:1092–1117

Keenan TM, Folch A (2008) Biomolecular gradients in cell culture systems. Lab Chip 8(1):34–57

Kim SR, Teixeira AI, Nealey PF et al (2002) Fabrication of polymeric substrates with well-defined nanometer-scale topography and tailored surface chemistry. Adv Mater 14(20):1468–1472

Kim P, Jeong HE, Khademhosseini A et al (2006) Fabrication of non-biofouling polyethylene glycol micro- and nanochannels by ultraviolet-assisted irreversible sealing. Lab Chip 6(11):1432–1437

Kim P, Kwon KW, Park MC et al (2008) Soft lithography for microfluidics: a review. Biochip J 2(1):1–11

Kim S, Kim HJ, Jeon NL (2010) Biological applications of microfluidic gradient devices. Integr Biol 2(11–12):584–603

Kirby BJ (2010) Micro-and nanoscale fluid mechanics: transport in microfluidic devices. Cambridge University Press, New York: Jul 26

Koch TM, Münster S, Bonakdar N et al (2012) 3D traction forces in cancer cell invasion. PLoS One 7(3):e33476

Kollmannsberger P, Mierke CT, Fabry B (2011) Nonlinear viscoelasticity of adherent cells is controlled by cytoskeletal tension. Soft Matter 7(7):3127–3132

Kramer N, Walzl A, Unger C et al (2013) In vitro cell migration and invasion assays. Mutat Res 752(1):10–24

Ladoux B (2009) Biophysics: cells guided on their journey. Nat Phys 5(6):377–378

Lange JR, Fabry B (2013) Cell and tissue mechanics in cell migration. Exp Cell Res 319(16):2418–2423

Lauffenburger DA, Horwitz AF (1996) Cell migration: a physically integrated molecular process. Cell 84(3):359–369

Le Roux-Goglin E, Varon C, Spuul P et al (2012) Helicobacter infection induces podosome assembly in primary hepatocytes in vitro. Eur J Cell Biol 91:161–170

Lee J, Park NC, Whitesides GM (2003) Solvent compatibility of poly (dimethylsiloxane)-based microfluidic devices. Anal Chem 75(23):6544–6554

Lee JH, Gu Y, Wang H et al (2012) Microfluidic 3D bone tissue model for high-throughput evaluation of wound-healing and infection-preventing biomaterials. Biomaterials 33(4):999–1006

Leichlé T, Lin YL, Chiang PC et al (2012) Biosensor-compatible encapsulation for pre-functionalized nanofluidic channels using asymmetric plasma treatment. Sensors Actuators B Chem 161(1):805–810

Li J, Francis L (2011) Microfluidic devices for studying chemotaxis and electrotaxis. Trends Cell Biol 21(8):489–497

Li L, Marchant RE, Dubnisheva A et al (2011) Anti-biofouling sulfobetaine polymer thin films on silicon and silicon nanopore membranes. J Biomater Sci Polym Ed 22(1–3):91–106

Li J, Zhu L, Zhang M et al (2012) Microfluidic device for studying cell migration in single or co-existing chemical gradients and electric fields. Biomicrofluidics 6(2):024121

Li R, Hebert JD, Lee TA et al (2017) Macrophage-secreted TNFα and TGFβ1 influence migration speed and persistence of cancer cells in 3D tissue culture via independent pathways. Cancer Res 77(2):279–290

Liao KT, Chou CF (2012) Nanoscale molecular traps and dams for ultrafast protein enrichment in high-conductivity buffers. J Am Chem Soc 134(21):8742–8745

Linder S (2007) The matrix corroded: podosomes and invadopodia in extracellular matrix degradation. Trends Cell Biol 17(3):107–117

Linder S, Wiesner C (2015) Tools of the trade: podosomes as multipurpose organelles of monocytic cells. Cell Mol Life Sci 72:121–135

Liu J, Zheng H, Poh PS et al (2015) Hydrogels for engineering of perfusable vascular networks. Int J Mol Sci 16(7):15997–16016

Lo CM, Wang HB, Dembo M et al (2000) Cell movement is guided by the rigidity of the substrate. Biophys J 79(1):144–152

Luxenburg C, Geblinger D, Klein E et al (2007) The architecture of the adhesive apparatus of cultured osteoclasts: from podosome formation to sealing zone assembly. PLoS One 2(1):e179

Manz A, Graber N, Widmer HÁ (1990) Miniaturized total chemical analysis systems: a novel concept for chemical sensing. Sensors Actuators B Chem 1(1–6):244–248

Mark D, Haeberle S, Roth G et al (2010) Microfluidic lab-on-a-chip platforms: requirements, characteristics and applications. Microfluidics based microsystems. Springer, Dordrecht, pp 305–376

Martinez AW, Phillips ST, Whitesides GM et al (2009) Diagnostics for the developing world: microfluidic paper-based analytical devices. Anal Chem 82(1):3–10

Masedunskas A, Milberg O, Porat-Shliom N et al (2012) Intravital microscopy: a practical guide on imaging intracellular structures in live animals. BioArchitecture 2(5):143–157

McCaig CD, Rajnicek AM, Song B et al (2005) Controlling cell behavior electrically: current views and future potential. Physiol Rev 85(3):943–978

McClelland R, Wauthier E, Uronis J et al (2008) Gradients in the liver's extracellular matrix chemistry from periportal to pericentral zones: influence on human hepatic progenitors. Tissue Eng Part A 14(1): 59–70

McDonald JC, Whitesides GM (2002) Poly(dimethyl siloxane) as a material for fabricating microfluidic devices. Acc Chem Res 35(7):491–499

McUsic AC, Lamba DA, Reh TA (2012) Guiding the morphogenesis of dissociated newborn mouse retinal cells and hES cell-derived retinal cells by soft lithography-patterned microchannel PLGA scaffolds. Biomaterials 33(5):1396–1405

Moreno-Arotzena O, Mendoza G, Cóndor M et al (2014) Inducing chemotactic and haptotactic cues in microfluidic devices for three-dimensional in vitro assays. Biomicrofluidics 8(6):064122

Nemir S, West JL (2010) Synthetic materials in the study of cell response to substrate rigidity. Ann Biomed Eng 38(1):2–20

Nge PN, Rogers CI, Woolley AT (2013) Advances in microfluidic materials, functions, integration, and applications. Chem Rev 113(4):2550–2583

Palchesko RN, Zhang L, Sun Y et al (2012) Development of polydimethylsiloxane substrates with tunable elastic modulus to study cell mechanobiology in muscle and nerve. PLoS One 7(12): e51499

Park J, Kim DH, Kim G et al (2010a) Simple haptotactic gradient generation within a triangular microfluidic channel. Lab Chip 10(16):2130–2138

Park JY, Yoo SJ, Lee EJ et al (2010b) Increased poly (dimethylsiloxane) stiffness improves viability and morphology of mouse fibroblast cells. Biochip J 4(3):230–236

Parsons JT, Horwitz AR, Schwartz MA (2010) Cell adhesion: integrating cytoskeletal dynamics and cellular tension. Nat Rev Mol Cell Biol 11(9):633

Paterson EK, Courtneidge S (2018) Invadosomes are coming: new insights into function and disease relevance. FEBS J 285:8–27

Pelham RJ Jr, Wang YL (1998) Cell locomotion and focal adhesions are regulated by the mechanical properties of the substrate. Biol Bull 194(3):348–350

Peng Z, Soper SA, Pingle MR et al (2010) Ligase detection reaction generation of reverse molecular beacons for near real-time analysis of bacterial pathogens using single-pair fluorescence resonance energy transfer and a cyclic olefin copolymer microfluidic chip. Anal Chem 82(23):9727–9735

Ren K et al (2011) Whole-Teflon microfluidic chips. Proc Natl Acad Sci 108(20):8162–8166

Ridley AJ, Schwartz MA, Burridge K et al (2003) Cell migration: integrating signals from front to back. Science 302(5651):1704–1709

Robinson KR, Kenneth R (1985) The responses of cells to electrical fields: a review. J Cell Biol 101(6):2023–2027

Roca-Cusachs P, Sunyer R, Trepat X (2013) Mechanical guidance of cell migration: lessons from chemotaxis. Curr Opin Cell Biol 25(5):543–549

Rogers CI, Pagaduan JV, Nordin GP et al (2011) Single-monomer formulation of polymerized polyethylene glycol diacrylate as a nonadsorptive material for microfluidics. Anal Chem 83(16):6418–6425

Rottiers P, Saltel F, Daubon T et al (2009) TGFβ-induced endothelial podosomes mediate basement membrane collagen degradation in arterial vessels. J Cell Sci 122:4311–4318

Sackmann EK, Fulton AL, Beebe DJ (2014) The present and future role of microfluidics in biomedical research. Nature 507(7491):181–189

Salomon S, Leichlé T, Nicu L (2011) A dielectrophoretic continuous flow sorter using integrated microelectrodes coupled to a channel constriction. Electrophoresis 32(12):1508–1514

Seano G, Primo L (2015) Podosomes and invadopodia: tools to breach vascular basement membrane. Cell Cycle 14(9):1370–1374

Sherman MA, Kennedy JP, Ely DL et al (1999) Novel polyisobutylene/polydimethylsiloxane bicomponent networks: III. Tissue compatibility. J Biomater Sci Polym Ed 10(3):259–269

Spuul P, Ciufici P, Veillat V et al (2014) Importance of RhoGTPases in formation, characteristics, and functions of invadosomes. Small GTPases 5:e28195

Spuul P, Chi P-Y, Billottet C et al (2016a) Microfluidic devices for the study of actin cytoskeleton in constricted environments: Evidence for podosome formation in endothelial cells exposed to a confined slit. Methods 94:65–74

Spuul P, Daubon T, Pitter B et al (2016b) VEGF-A/Notch-induced podosomes proteolyse basement membrane collagen-IV during retinal sprouting angiogenesis VEGF/Notch-induced podosomes mediate basement membrane collagen-IV proteolysis during sprouting angiogenesis in vivo. Cell Rep 17(2):484–500

Sriram KK, Yeh JW, Lin YL et al (2014) Direct optical mapping of transcription factor binding sites on field-stretched λ-DNA in nanofluidic devices. Nucleic Acids Res 42:e85

Sriram KK, Nayak S, Pengel S et al (2017) 10 nm deep, sub-nanoliter fluidic nanochannels on germanium

for attenuated total reflection infrared (ATR-IR) spectroscopy. Analyst 142(2):273–278
Stolz M, Raiteri R, Daniels AU et al (2004) Dynamic elastic modulus of porcine articular cartilage determined at two different levels of tissue organization by indentation-type atomic force microscopy. Biophys J 86(5):3269–3283
Su YH, Chiang PC, Cheng LJ et al (2015) High aspect ratio nanoimprinted grooves of poly(lactic-co-glycolic acid) control the length and direction of retraction fibers during fibroblast cell division. Biointerphases 10(4):041008
Swami N, Chou CF, Ramamurthy V et al (2009) Enhancing DNA hybridization kinetics through constriction-based dielectrophoresis. Lab Chip 9(22):3212–3220
Tabeling P (2005) Introduction to microfluidics. Oxford University Press, Oxford
Tambe DT, Hardin CC, Angelini TE et al (2011) Collective cell guidance by cooperative intercellular forces. Nat Mater 10(6):469
Tan JL, Tien J, Pirone DM et al (2003) Cells lying on a bed of microneedles: an approach to isolate mechanical force. Proc Natl Acad Sci 100(4):1484–1489
Tan SH, Nguyen NT, Chua YC et al (2010) Oxygen plasma treatment for reducing hydrophobicity of a sealed polydimethylsiloxane microchannel. Biomicrofluidics 4(3):032204
Teerapanich P, Pugnière M, Henriquet C, Lin YL, Naillona A, Josepha P, Chou CF, Leichle T (2018) Nanofluidic fluorescence microscopy with integrated concentration gradient generation for one-shot parallel kinetic assays. Sensors Actuators B 274:338–342
Trepat X, Fredberg JJ (2011) Plithotaxis and emergent dynamics in collective cellular migration. Trends Cell Biol 21(11):638–646
Trepat X, Wasserman MR, Angelini TE et al (2009) Physical forces during collective cell migration. Nat Phys 5(6):426–430
van den Dries K, van Helden SF, re Riet J et al (2012) Geometry sensing by dendritic cells dictates spatial organization and PGE(2)-induced dissolution of podosomes. Cell Mol Life Sci 69(11):1889–1901
van den Dries K, Meddens MB, de Keijzer S et al (2013) Interplay between myosin IIA-mediated contractility and actin network integrity orchestrates podosome composition and oscillations. Nat Commun 4:1412
Van Haastert PJ, Devreotes PN (2004) Chemotaxis: signalling the way forward. Nat Rev Mol Cell Biol 5(8):626
Varon C, Tatin F, Moreau V et al (2006) Transforming growth factor beta induces rosettes of podosomes in primary aortic endothelial cells. Mol Cell Biol 26(9):3582–3594
Veillat V, Spuul P, Daubon T, Egaña I, Kramer I et al (2015) Podosomes: multipurpose organelles? Int J Biochem Cell Biol Organelles Focus 65:52–60
Velve-Casquillas G, Le Berre M, Piel M et al (2010) Microfluidic tools for cell biological research. Nano Today 5(1):28–47
Vogel V, Sheetz M (2006) Local force and geometry sensing regulate cell functions. Nat Rev Mol Cell Biol 7(4):265–275
Wang N, Tolić-Nørrelykke IM, Chen J et al (2002) Cell prestress. I. Stiffness and prestress are closely associated in adherent contractile cells. Am J Physiol Cell Physiol 282(3):C606–C616
Wang CC, Kao YC, Chi PY et al (2011) Asymmetric cancer-cell filopodium growth induced by electric-fields in a microfluidic culture chip. Lab Chip 11(4):695–699
Wang S, Li E, Gao Y et al (2013) Study on invadopodia formation for lung carcinoma invasion with a microfluidic 3D culture device. PLoS One 8(2):e56448
Wheeler AR, Throndset WR, Whelan RJ et al (2003) Microfluidic device for single-cell analysis. Anal Chem 75(14):3581–3586
Wolf K, Te Lindert M, Krause M et al (2013) Physical limits of cell migration: control by ECM space and nuclear deformation and tuning by proteolysis and traction force. J Cell Biol 201(7):1069–1084
Wu J, Wu X, Lin F (2013) Recent developments in microfluidics-based chemotaxis studies. Lab Chip 13(13):2484–2499
Yeh JW, Taloni A, Chen YL et al (2012) Entropy-driven single molecule tug-of-war of DNA at micro-nanofluidic interfaces. Nano Lett 12(3):1597–1602
Yeo LY, Chang HC, Chan PP et al (2011) Microfluidic devices for bioapplications. Small 7(1):12–48
Yin H, Marshall D (2012) Microfluidics for single cell analysis. Curr Opin Biotechnol 23(1):110–119
Yoshida A, Kanno H, Watabe D et al (2008) The role of heparin-binding EGF-like growth factor and amphiregulin in the epidermal proliferation of psoriasis in cooperation with TNFα. Arch Dermatol Res 300(1):37–45
Young EW, Beebe DJ (2010) Fundamentals of microfluidic cell culture in controlled microenvironments. Chem Soc Rev 39(3):1036–1048
Yu CH, Groves JT (2010) Engineering supported membranes for cell biology. Med Biol Eng Comput 48(10):955–963
Yu CH, Rafig NB, Krishnasamy A et al (2013) Integrin-matrix clusters form podosome-like adhesions in the absence of traction forces. Cell Rep 5(5):1456–1468
Zare RN, Kim S (2010) Microfluidic platforms for single-cell analysis. Annu Rev Biomed Eng 12:187–201
Zhang Y, He Y, Bharadwaj S et al (2009) Tissue-specific extracellular matrix coatings for the promotion of cell proliferation and maintenance of cell phenotype. Biomaterials 30(23–24):4021–4028

Collective Cell Migration in Development

7

Linus Schumacher

Abstract

Collective cell migration is a key process in developmental biology, facilitating the bulk movement of cells in the morphogenesis of animal tissues. Predictive understanding in this field remains challenging due to the complexity of many interacting cells, their signalling, and microenvironmental factors – all of which can give rise to non-intuitive emergent behaviours. In this chapter we discuss biological examples of collective cell migration from a range of model systems, developmental stages, and spatial scales: border cell migration and haemocyte dispersal in *Drosophila*, gastrulation, neural crest migration, lateral line formation in zebrafish, and branching morphogenesis; as well as examples of developmental defects and similarities to metastatic invasion in cancer. These examples will be used to illustrate principles that we propose to be important: heterogeneity of cell states, substrate-free migration, contact-inhibition of locomotion, confinement and repulsive cues, cell-induced (or self-generated) gradients, stochastic group decisions, tissue mechanics, and reprogramming of cell behaviours. Understanding how such principles play a common, overarching role across multiple biological systems may lead towards a more integrative understanding of the causes and function of collective cell migration in developmental biology, and to potential strategies for the repair of developmental defects, the prevention and control of cancer, and advances in tissue engineering.

Keywords

Cell migration · Developmental biology · Collective behaviour · Morphogenesis · Cell interactions

7.1 Introduction

In this chapter, we will introduce the reader to selected model systems for collective cell migration. We deliberately choose examples at various stages of animal development, in a range of organisms, spanning multiple time-scales and cell population sizes. In each case, we will motivate the use of the model system, and describe what is known about the mechanism of collective migration in that system. To conclude each section, we single out a principle of collective cell migration that a particular system provides insight to or is promising to do so. In many cases,

L. Schumacher (✉)
MRC Centre for Regenerative Medicine, University of Edinburgh, Edinburgh, UK
e-mail: Linus.Schumacher@ed.ac.uk

C. A. M. La Porta, S. Zapperi (eds.), *Cell Migrations: Causes and Functions*, Advances in Experimental Medicine and Biology 1146,
https://doi.org/10.1007/978-3-030-17593-1_7

a particular biological example could illustrate multiple principles, and our choice is not unique. Heterogeneity of cell states could be exemplified by neural crest as well as border cell migration, and contact-inhibition of locomotion by *Xenopus* neural crest as well as *Drosophila* haemocyte dispersal, to give just two examples. Our choice of model systems is by no means exhaustive. We have mainly focused on in vivo examples. In vitro systems have undoubtedly contributed to our understanding of the mechanisms of collective cell migration under controlled experimental conditions, and reviews can be found elsewhere (Ladoux and Mège 2017; Trepat and Sahai 2018). Here we hope to provide vignettes that together form more than the sum of parts and provide the reader with an emergent appreciation of collective cell migration in development.

7.2 Border Cell Migration

Drosophila border cell migration (Inaki et al. 2012) could be described as the hydrogen atom of collective cell migration. Consisting of a handful of cells, it serves as a minimal example in which the migration of the group differs from that of the individual, i.e., a "collection of cells moving together and affecting one another while doing so" (Rørth 2012, which forms our working definition of collective cell migration). And just as the study of the hydrogen atom, the simplest atom, has advanced theoretical understanding in atomic physics, we stand to learn from focussing on minimal systems of collective cell migration.

In the formation of the *Drosophila* egg, a cluster of about eight border cells migrates across the nurse cells from the anterior of the egg chamber to the oocyte on the posterior. This journey covers a distance of about 200 μm (Prasad et al. 2011), at about 0.5 μm/min (Montell et al. 2012). The cluster of cells goes on to form an egg shell structure that enables sperm entry, so their positioning at the oocyte is important for egg fertilization (Montell et al. 2012). The group consists of migrating border cells and non-migrating polar cells, and exhibits both leading/trailing polarity, meaning that the cells at the front and back of the group look and act differently, as well as inner/outer polarity, meaning that the polar and border cells are different (Montell et al. 2012). During migration, frequent reorientations of the cluster occur, changing which cell is in the leading position (Prasad and Montell 2007).

Border cell migration follows guidance signals present in their microenvironment, like many of the examples of collective cell migration that we will encounter in this chapter. These signals include the attractant Pvf1, which is read via the receptor tyrosine kinase PDGF/VEGF receptor. Leading and trailing cells show differences in the activity of this receptor (Janssens et al. 2010). This heterogeneity between cells in responding to guidance cues thus imparts directionality at the group level (Inaki et al. 2012), with the group being led by the cell with high activity of the receptor for the guidance cue. This shows that the cells are indeed acting as a *collective*, moving differently than each cell undergoing its own guided migration.

7.2.1 Heterogeneity of Cell States

Differences in cells' states, and thus their migratory behaviour, are an important aspect of cell populations that can affect their collective migration. *Drosophila* border cells provide a clear example of leader-follower heterogeneity between cells in a migrating group, a form of heterogeneity frequently studied in collective cell migration. The dynamic nature of the leader cell states with frequent changeover between cells highlights that this heterogeneity can emerge from interaction between cells and the environment, and need not be pre-specified. And even though the different cell states are not fixed, and may in some cases lie on a continuum (Schumacher 2019), they have turned out to be crucial to understanding the mechanism of group migration in this system. This is often misunderstood in debates about whether leader and follower cells exist – that is beside the point, the question is whether the concept provides a useful description. This is nicely summarised in a quote that bears repeating: "leader and follower cells should be considered as different cell states and not different cell types" (Rørth 2012).

7.3 Gastrulation

Gastrulation is the earliest and one of the most important examples of collective cell movement in the development of an animal embryo. In terms of relative cell numbers, it is also the largest remodelling of tissue structure – involving most cells at this stage of development to some degree. From an initially homogenous seeming mass of cells in the early embryo, an extensive rearrangement of cells establishes the three tissue layers ectoderm (giving rise to epidermis and nerves), mesoderm (turning into connective tissue, muscle, skeleton, etc.), and endoderm (giving rise to epithelial linings), broadly speaking the outer, middle, and inner tissue types. Gastrulation has thus been termed, and often quoted, as the "most important time in your life" (Wolpert 2008).

The details of the choreography of cell movements during gastrulation differ in their details in different species. The intricacies of these differences have been thoroughly documented elsewhere (see for example (Stern), or Keller (2005), and for a physics perspective, Forgacs and Newman 2005). Here we are restricting ourselves to avian and mouse gastrulation, and wish to only convey a general sense of the course of events: Mesoderm and endoderm precursors migrate inwards into the embryo, establishing the three tissue layers together with the ectoderm (Gilbert). The types of movement that cells undergo during gastrulation, and which global tissue deformations these produce, differ in different organisms. What they have in common is that the process of gastrulation turns an embryo from a relatively unstructured clump of cells into a layered tissue, with a head-to-tail body axis, and a distinct "outside" and "inside" that will go on to form the gut and respiratory system.

7.3.1 Collective Cell Migration Without a Substrate?

During gastrulation the different tissue layer precursors move with respect to each other, but in the absence of a substrate to move on or through. In other words, there does not seem to be an absolute coordinate system, unlike cases of cell migration usually considered. This is a vivid demonstration that collective cell migration can occur without an external substrate, but, in a sense, with other cells acting as the substrate. One could argue that this situation is not so different in cell migration within a tissue, but here the distinction between moving cells and resident tissue is usually clearer. This is reinforced through the presence of extracellular matrix (ECM), which provides a passive medium for cells to move through and interact with (though ECM may also play a role during early gastrulation, see Latimer and Jessen 2010). In gastrulation, one could distinguish between cells that are actively migrating or changing their shape, and those undergoing passive rearrangement in response to intercellular forces. Methods to quantify the contributions of cell shape changes and rearrangements are an active field of current research (Blanchard 2017; Dicko et al. 2017; Firmino et al. 2016; Lye et al. 2015; Rozbicki et al. 2015).

7.4 Haemocyte Dispersal

Haemocyte dispersal provides an example of multicellular migration that is not densely packed, but in which the migration is nonetheless influenced by the interactions of cells between each other. It thus provides an important sample on the spectrum of collective cell migration (Schumacher et al. 2016). *Drosophila* haemocytes spread out in the embryo during development, originating from the head mesoderm (Tepass et al. 1994). They are required for a functioning immune response and thus broadly similar to macrophages. *Drosophila* haemocytes migrate as single cells, but collectively need to arrange in an evenly spread, lattice-like pattern (Davis et al. 2012).

In vivo tracking of haemocyte movement (Davis et al. 2012) revealed that cells accelerate away from each other after encounters. These "collisions" take on the order of a few minutes, during which the cells were observed to extend microtubule-driven protrusions towards each other, make contact, and then retract. This movement could be described by a persistent random walk, with an additional "contact-inhibition of locomotion" interaction that induces displacements away from nearby cells. By varying the strength of this interaction, Davis et al. (2012) could simulate the effect of haemocyte dispersal with and without repulsive collisions. Simulated dispersal without repulsive collisions, or with only a volume exclusion interaction, failed to produce the regular patterning of cell positions observed in the embryo. Further experiments with *diaphanous* mutants, showing uncoordinated cell-cell repulsion, confirmed that this led to break-down of ordered pattern formation in vivo (Davis et al. 2015).

7.4.1 Contact-Inhibition of Locomotion

Contact-inhibition of locomotion had, for a long time, been primarily observed in vitro (Abercrombie and Heaysman 1953, 1954; Loeb 1921). In recent years several studies have argued for its relevance for pattern formation in embryonal development. In the case of *Drosophila* haemocyte dispersal, this has been demonstrated through detailed in vivo imaging, genetic perturbation, and computational simulations. In *Xenopus* cephalic neural crest, in vivo studies (Carmona-Fontaine et al. 2008), aided by in vitro experiments and again by computational modeling, have also pointed to a role for contact-inhibition of locomotion, coupled with co-attraction, as a mechanism to promote collective cell migration. The importance of contact-inhibition of locomotion has been called into doubt in chick cranial neural crest (Genuth et al. 2018), so it remains unclear how relevant this mechanism is in the neural crest generally. One possibility is that interactions between cells lie on a continuum ranging from contact guidance and volume exclusion to the repulsive contact-inhibition of locomotion described above (Schumacher et al. 2016). Formulating integrative models that offer such unifying descriptions of the mechanisms of collective cell migration is a subject of future work (see also Sect. 7.5).

7.5 Neural Crest

The migration of the neural crest is one of the most striking and versatile examples of collective cell migration in developmental biology. Neural crest cells are a migratory cell population found in the vertebrate embryo that develops into a range of tissues throughout the body, such as peripheral nerves and smooth muscle, as well as contributing to many others, such as heart and bone (Kulesa et al. 2010; Le Douarin 2004). They originate from the dorsal neural tube, which develops into the brain and spinal cord, undergo EMT and migrate over distances of up to 1 mm through the mesoderm of the growing embryo, first lateral and then ventral. Neural crest in different organisms and different body parts exhibit a range of migration morphologies, and therefore offer a system to investigate how the common mechanisms may play a role in diverse biological settings. As such, they have become a popular model organism for long-range mesenchymal collective cell migration.

Developmental defects associated with failed or incomplete migration are known as neurocristopathies (Benish 1975) and include pigmentation defects, cleft lip, cleft palate, and incomplete innervation of the gut (Hirschsprung's disease) (Lake and Heuckeroth 2013). The invasive nature of their migration, and the fact that some cancers, such as neuroblastoma and melanoma, derive from the neural crest, have attracted attention to this system for the study of metastatic invasion. We will discuss these aspects further in Sects. 7.8 and 7.9.

Mechanisms of neural crest migration appear as diverse as the vertebrate organisms they have been studied in. The migrating neural crest forms discrete streams along the head-tail axis of the body (Kulesa and Gammill 2010). These streams are sculpted by a combination of inhibitory and repulsive factors, as well as other tissue structures serving as barriers (e.g. the otic vesicle, which forms the inner ear). Common features exist across the different model systems (Cebra-Thomas et al. 2013; Krotoski et al. 1988; Löfberg et al. 1980; Nikitina et al. 2009; Reyes et al. 2010; Schilling and Kimmel 1994; Serbedzija et al. 1989, 1992), but with important differences between organisms as well as between different positions along the head-tail axis. Cells often follow guidance factors, such as VEGF in chick cranial migration (McLennan et al. 2010, 2015b) and SDF1 in Xenopus cephalic migration (Theveneau et al. 2010). Neural crest cells migrate towards target zones, such as the branchial arches, where they proliferate and differentiate (Ridenour et al. 2014), but also need to be distributed along the migratory route and can undergo secondary migration at later times, such as in the formation of vertebral sympathetic ganglia from trunk neural crest (Kasemeier-Kulesa et al. 2015). Interactions between cells are important for proper group migration (McKinney et al. 2011; Teddy and Kulesa 2004), including follow-the-leader migration in chick (McLennan et al. 2012, 2015a) and cell-cell attraction (Carmona-Fontaine et al. 2011) with contact-inhibition-of-locomotion in Xenopus (Carmona-Fontaine et al. 2008) (but not in chick, see Genuth et al. 2018). It remains unresolved whether the diversity of behaviours displayed by the large number of neural crest systems can be reconciled by a universal set of mechanisms.

7.5.1 Confinement and Repulsive Cues

Since the neural crest is such a diverse and popular model system for collective cell migration in development, it would be short-sighted to highlight just one principle as important. Nevertheless, in balance with the other sections of this chapter, let us highlight the remarkable organisation of neural crest cells migration in long streams (on the order of 1 mm). What generates the required cohesion and persistence is incompletely understood, but a few pieces of the puzzle have been uncovered. We already mentioned the role of inhibitory/repulsive cues (ephrins, semaphorins) to shape the cells delaminating from the neural tube into discrete streams (Kulesa and Gammill 2010). Recent studies have further suggested new mechanisms for confining cells through versican (in *Xenopus*, Szabó et al. 2016) and restricting their invasion through DAN (in chick, McLennan et al. 2017). In Xenopus cranial neural crest, recent work calls into question the importance of guidance cues in early stream formation, and instead proposes that these neural crest streams initially emerge from "on short-range repulsion and asymmetric attraction between neighboring tissues" (Szabó et al. 2019). In avian neural crest, and in collective cell migration more generally, repulsive cues likely remain an important tool for tissues to control and confine collective cell migration, and can be found in other systems, such as the zebrafish germline (Paksa et al. 2016).

7.6 Lateral Line Formation

The lateral line is a system of mechanosensory organs in aquatic vertebrates, and its development is commonly studied in zebrafish (Haas and Gilmour 2006). Its formation is an example of epithelial collective cell migration, in which a cohesive group of cells, about 100 μm in length, migrates over millimeters in the growing zebrafish embryo. Unlike most examples discussed in this chapter, the migration is effectively one-dimensional, offering a simplified perspective on directional symmetry breaking in a system of interacting cells.

The lateral line primordium (LLP) is a cohesive group of on the order of 100 cells that migrate collectively along the side of the zebrafish embryo and form multicellular structures in their wake that later make up the lateral line (Haas and Gilmour 2006). They migrate along a strip of chemoattractant Cxcl12/Sdf1. This ligand is not expressed in a gradient, however, it is only in interaction with the migrating group of cells that directionality is established (Streichan et al. 2011). Leading cells at the front of the LLP read out the chemoattractant via the receptor Cxcr4, while trailing cells sequester the ligand via Cxcr7 to create a local gradient, and both are required for successful migration (Donà et al. 2013). As the LLP migrates, subgroups of cells organise into rosette-like structures via adherens junctions (Revenu et al. 2014). These multicellular structures have a luminal space at their core, which is thought to enable coordination of cells in the group via local signalling (Durdu et al. 2014). The rosettes split from the migrating group and stay behind to form the aforementioned sensory organs.

7.6.1 Cell-Induced or Self-Generated Gradients

An alternative to the migration along pre-established gradients of morphogens or chemoattractants is the dynamic creation and interpretation of local signalling gradients. In an otherwise uniform concentration of chemoattractant, groups of cells can locally create gradients by internalising or breaking down the signal in their vicinity. These self-generated or cell-induced gradients have been investigated in a number of systems, such as the neural crest (Kulesa et al. 2010; McLennan et al. 2012; Schumacher 2019) *Dictyostelium* (Tweedy et al. 2016), and melanoma cells (Muinonen-Martin et al. 2014), but in the context of development they are probably best understood in lateral line migration (Donà et al. 2013; Streichan et al. 2011). Collective migration in self-generated gradients is conceptually similar to aggregation with self-secreted attractive signals, which have been studied mathematically in some of the earliest models of chemotaxis (Keller and Segel 1970a,b). Cell-induced gradients may be an environment where leader-follower heterogeneity (see Sec. 7.2.1) is advantageous, depending on the kinetics of gradient formation, as been explored in recent theoretical studies (Hopkins and Camley 2019; Schumacher 2019).

7.7 Branching Morphogenesis

The tree-like structures produced by branching morphogenesis appear both beautifully complex, and also self-similar at multiple scales, or fractal-like (Iber and Menshykau 2013). This makes them potentially amenable to production through simple developmental programs, and thus branching morphogenesis has been of long-standing interest to developmental biologists and mathematical biologists (Iber and Menshykau 2013; Murray et al. 1983). Examples include lung (and trachea in insects), kidney, pancreas, blood vessels, prostate, salivary and mammary glands. It encompasses the growth of tree-like ductal networks, thus achieving a high surface area to exchange molecules, such as oxygen, or metabolic products, with the environment or other tissues. While much of the biological research in the past decades has focussed on the molecular (and also mechanical) control of branching and elongation, we want to consider it here as an example of collective cell behavior with an emergent, large-scale structure.

Branching and annihilating random walks (BARWs) have recently been put forward as a promising candidate for a unified theory of branching morphogenesis (Hannezo et al. 2017), explaining statistical patterns of the network structures in mouse mammary gland, kidney, and humane prostate. In branching random walks, ducts elongate and branch stochastically, while in this particular variant growth of tips is terminated when they contact existing ducts (in addition to branching there can also be budding from the side of existing ducts, which plays a role for example in early lung formation, see Iber and Menshykau 2013). A minimal BARW model was able to reproduce statistics such as the distribution of subtree sizes in several organs, with only experimentally determined parameters (Hannezo et al. 2017).

Within this general framework, the molecular control of branching and annihilation events may be tissue-specific: In mouse mammary gland, the tip termination can be induced by implanted sources of TGF-β1 (Hannezo et al. 2017), and branching is promoted by FGF10 (Hannezo et al. 2017). In mouse kidney, the TGF-β-related BMP7 has been implicated in tip termination (Davies et al. 2014), while proliferation is driven by GDNF (Lambert et al. 2017).

While the BARW model is remarkably successful at reproducing global statistical features of the tree structures, small modifications have been necessary to match the detailed features of some particular tissues. For example, the relatively ordered three dimensional structure of kidney ducts was more faithfully reproduced by a BARW with additional self-repulsive interactions of the growing tips. Then again, a more complex model can always better describe existing data than a simpler one. The minimal BARW model is an attractive paradigm for branching morphogenesis precisely due to its simplicity.

7.7.1 Stochastic Group Decisions

The use of BARW models for branching morphogenesis nicely illustrates how seemingly complex and (statistically) stereotypic structures can form through stochastic "decisions". This occurs at two levels: the overall organ structure arises from the interplay of many stochastic branching events, and the individual branching event is itself a stochastic event in which many cells have to coordinate. The means by which a group of cells in an individual tip conduct a poll or otherwise decide whether to elongate, branch, or terminate, and do so in a seemingly stochastic manner, remain hitherto unresolved. This exemplifies a common challenge in the pursuit of quantitative understanding of collective cell movements, namely phenomena that occur at the mesoscale between the cell- and tissue-levels (Blanchard et al. 2018).

7.8 Developmental Defects

Developmental defects arise when developmental processes go awry. In the context of collective cell migration, this can occur when the migration is mistargeted, mistimed, or miscoordinated. Each of the sections in this chapter would deserve its own discussion of associated developmental defects, but here we will once more focus on the neural crest and the aforementioned neurocristopathies (Benish 1975) (developmental defects that are related to failures in neural crest cell migration). From the many neurocristopathies we pick an illustrative example from enteric neural crest migration.

In healthy embryonic development, enteric neural crest cells colonise the growing gut through migration and proliferation, and this is important for innervation of the gut, i.e., the development of the enteric nervous system. The neurocristopathy known as Hirschsprung's disease affects about 1 in 5000 live births (Lake and Heuckeroth 2013). It can have multiple causes, and one of its symptoms is failed innervation of parts of the gut, which can lead to life-threatening obstruction of the bowels (Lake and Heuckeroth 2013). Understanding the causes of failed enteric nervous system development in Hirschprung's disease could lead to therapeutic strategies to prevent or repair this developmental defect.

In experiments with chick enteric neural crest, it was found that stiffening of the gut mesenchyme through externally applied stretch prevents normal colonisation of the gut (Chevalier et al. 2016). This was further supported by experiments in which enteric neural crest were embedded in 3D gels, and found to invade less far into stiffer 3D gels than they migrated in more compliant ones (Chevalier et al. 2016). Stiffening of the tissue is part of the normal developmental process, but, as the described work shows, mistiming of this process, e.g. if the migration of neural crest cells is delayed, can lead to failed innervation. This suggests a possible cause for the symptoms of Hirschsprung's disease. Furthermore, it further highlights (one of several) challenges faced by potential treatments of failed gut colonisation: If migratory neural crest cells are transplanted later in development, they may not be able to migrate and colonise effectively in the developed, stiffened gut. On the other hand, it may point the way for future research how to modify the transplanted cells and/or the tissue microenvironment to enable repair of the developmental defect.

7.8.1 Cell Migration and Substrate Mechanics... It's Complicated

Changes in the mechanical properties of the ECM and surrounding tissues can affect the migration of cells in different ways. In the example above we have seen an inhibition of invasive migration through stiffening of the substrate tissue. In contrast to this, in *Xenopus* cephalic neural crest, stiffening of the mesoderm tissue in contact with the neural crest cells (Barriga et al. 2018) triggers the start of their migration.

The different effects of tissue stiffening in chick enteric and *Xenopus* cephalic neural crest could have a number of reasons. One difference is the magnitude of the elastic modulus of the tissue in question, which is an order of magnitude higher in the chick gut (Chevalier et al. 2016) than in the *Xenopus* head (Barriga et al. 2018). It is reasonable to consider that the relationship between cell migration and substrate stiffness is non-monotonic, so that some stiffness is needed for migration, but too stiff a substrate hinders invasion. There is another difference between these two experimental systems: the dimensionality of the problem is different. Enteric neural crest cells have to migrate through the tissue that is stiffening (a 3D substrate), whereas in the cephalic neural crest it is the adjacent mesenchyme that stiffens, which forms a 2D contact with the group of cells. Further research will be needed to disentangle the different effects of substrate mechanics on collective cell migration in two- and three-dimensional environments. To summarise, how changes in mechanics of a substrate tissue affect migration of a cell collective can depend on a number of factors, including timing, magnitude, and dimensionality.

7.9 Metastatic Invasion

Many aspects of collective cell migration in development are also found in metastatic invasion of cancer cells (Maguire et al. 2015). Metastases are the prime reason why cancers are lethal. As cancerous cells spread and nest secondary tumours throughout the body, our ability to surgically remove or target them with radiotherapy diminishes. Understanding what makes cancer cells migrate, and what enables them to invade healthy tissues, offers the prospect of controlling these misregulated collective cell behaviours. An introduction into mechanical factors of collective cancer cell migration and metastasis can be found in La Porta and Zapperi (2017, Chapter 7).

Cancer may, in part, be a reversion to embryonic development programs that suddenly become harmful when played out in the wrong time and place. The ability of embryonic cells to migrate and proliferate then becomes "a liability by contributing to tumorigenesis and metastasis" (Maguire et al. 2015). One example is, again, the neural crest, which as a lineage is the origin of melanoma, neuroblastoma and others cancers (Maguire et al. 2015), and whose invasive migration in embryonic development bears characteristics of metastatic cancer invasion. Coupled with the relative ease of transplantation in the chick embryo system, the neural crest and its embryonic microenvironment are a useful model system to study cancer metastasis in vivo (Bailey et al. 2012).

7.9.1 Reprogramming

When metastatic melanoma cells are transplanted into neural crest microenvironment, they migrate along normal neural crest migratory paths to target tissues without forming tumors (Hendrix et al. 2007; Kulesa et al. 2006). These results provide a tantalising possibility for anti-metastatic therapy: the embryonic microenvironmental signals could be exploited to reprogram cancer cells into a less harmful state (Kasemeier-Kulesa et al. 2018), and to directly constrain their invasive migration (McLennan et al. 2017). In addition to embryonic signals controlling collective cell migration providing preliminary candidates for cancer drugs, systems like the melanoma-chick transplant model also offer a cheap way to initially screen drugs for their anti-metastatic efficacy in vivo (Maguire et al. 2015).

7.10 Conclusion

In this chapter we have provided a brief overview of several examples of collective cell migration in development. The intent was to give the reader a broad selection of different biological systems, each with their own merits and fascinating problems to study. The selection has been necessarily biased towards the author's interest, and other reviews on the topic will provide different perspectives (Scarpa and Mayor 2016; Weijer 2009). An emerging trend that can be gleamed from the research discussed here, and hopefully throughout this book, is the integration of mathematical and computational models alongside experiments to interrogate the causes and function of cell migration with multidisciplinary approaches (Blanchard et al. 2018; Schumacher et al. 2016).

We have deliberately held back on quoting reams of results on molecular mechanisms, which can be found within the references cited in this chapter. Instead, we have opted to propose "principles", or, to phrase it more modestly, "themes for discussion" that link particular biological examples with concepts that may (or may not) help to move towards an overarching understanding of collective cell migration in developmental biology, and beyond. We hope that the reader will disagree with at least some of these, and that this disagreement may spark insightful discussion and further research.

References

Abercrombie M, Heaysman JE (1953) Observations on the social behaviour of cells in tissue culture: I. Speed of movement of chick heart fibroblasts in relation to their mutual contacts. Exp Cell Res 5:111–131. https://doi.org/10.1016/0014-4827(53)90098-6

Abercrombie M, Heaysman JE (1954) Observations on the social behaviour of cells in tissue culture: II. Monolayering of fibroblasts. Exp Cell Res 6:293–306. https://doi.org/10.1016/0014-4827(54)90176-7

Bailey CM, Morrison JA, Kulesa PM (2012) Melanoma revives an embryonic migration program to promote plasticity and invasion. Pigment Cell Melanoma Res 25:573–583. https://doi.org/10.1111/j.1755-148X.2011.01025.x

Barriga EH, Franze K, Charras G, Mayor R (2018) Tissue stiffening coordinates morphogenesis by triggering collective cell migration in vivo. Nature 554:523–527. https://doi.org/10.1038/nature25742

Benish BM (1975) The neurocristopathies: a unifying concept of disease arising in neural crest maldevelopment. Hum Pathol 6:128

Blanchard GB (2017) Taking the strain: quantifying the contributions of all cell behaviours to changes in epithelial shape. Philos Trans R Soc B 372:20150513

Blanchard GB, Fletcher AG, Schumacher LJ (2018) The devil is in the mesoscale: mechanical and behavioural heterogeneity in collective cell movement. Semin Cell Dev Biol. https://doi.org/10.1016/j.semcdb.2018.06.003

Carmona-Fontaine C, Matthews HK, Kuriyama S, Moreno M, Dunn GA, Parsons M, Stern CD, Mayor R (2008) Contact inhibition of locomotion in vivo controls neural crest directional migration. Nature 456:957–961. https://doi.org/10.1038/nature07441

Carmona-Fontaine C, Theveneau E, Tzekou A, Tada M, Woods M, Page KM, Parsons M, Lambris JD, Mayor R (2011) Complement fragment C3a controls mutual cell attraction during collective cell migration. Dev Cell 21:1026–1037. https://doi.org/10.1016/j.devcel.2011.10.012

Cebra-Thomas JA, Terrell A, Branyan K, Shah S, Rice R, Gyi L, Yin M, Hu Y, Mangat G, Simonet J, Betters E, Gilbert SF (2013) Late-emigrating trunk neural crest cells in turtle embryos generate an osteogenic ectomesenchyme in the plastron. Dev Dyn 242:1223–1235. https://doi.org/10.1002/dvdy.24018

Chevalier N, Gazquez E, Bidault L, Guilbert T, Vias C, Vian E, Watanabe Y, Muller L, Germain S, Bondurand N, Dufour S, Fleury V (2016) How tissue mechanical properties affect enteric neural crest cell migration. Sci Rep 6:20927. https://doi.org/10.1038/srep20927

Davies JA, Hohenstein P, Chang CH, Berry R (2014) A self-avoidance mechanism in patterning of the urinary collecting duct tree. BMC Dev Biol 14:1–12. https://doi.org/10.1186/s12861-014-0035-8

Davis JR, Huang C-Y, Zanet J, Harrison S, Rosten E, Cox S, Soong DY, Dunn GA, Stramer BM (2012) Emergence of embryonic pattern through contact inhibition of locomotion. Development 139:4555–4560. https://doi.org/10.1242/dev.082248

Davis JR, Luchici A, Mosis F, Thackery J, Salazar JA, Mao Y, Dunn GA, Betz T, Miodownik M, Stramer BM (2015) Inter-cellular forces orchestrate contact inhibition of locomotion. Cell 161:361–373. https://doi.org/10.1016/j.cell.2015.02.015

Dicko M, Saramito P, Blanchard GB, Lye CM, Sanson B, Étienne J (2017) Geometry can provide long-range mechanical guidance for embryogenesis. PLoS Comput Biol 13:1–30. https://doi.org/10.1371/journal.pcbi.1005443

Donà E, Barry JD, Valentin G, Quirin C, Khmelinskii A, Kunze A, Durdu S, Newton LR, Fernandez-Minan A, Huber W, Knop M, Gilmour D (2013) Directional

tissue migration through a self-generated chemokine gradient. Nature 503:285–289. https://doi.org/10.1038/nature12635
Durdu S, Iskar M, Revenu CC, Schieber N, Kunze A, Bork P, Schwab Y, Gilmour D (2014) Luminal signalling links cell communication to tissue architecture during organogenesis. Nature 515:120. https://doi.org/10.1038/nature13852
Firmino J, Rocancourt D, Saadaoui M, Moreau C, Gros J (2016) Cell division drives epithelial cell rearrangements during gastrulation in chick. Dev Cell 36:249–261. https://doi.org/10.1016/j.devcel.2016.01.007
Forgacs G, Newman SA (2005) Biological physics of the developing embryo. ISBN:9780511755576
Genuth MA, Allen CD, Mikawa T, Weiner OD (2018) Chick cranial neural crest cells use progressive polarity refinement, not contact inhibition of locomotion, to guide their migration. Dev Biol. https://doi.org/10.1016/j.ydbio.2018.02.016
Gilbert SF, Barresi MJF (2017) Developmental biology. ISBN: 9781605357386
Haas P, Gilmour D (2006) Chemokine signaling mediates self-organizing tissue migration in the Zebrafish lateral line. Dev Cell 10:673–680. https://doi.org/10.1016/j.devcel.2006.02.019
Hannezo E, Scheele CL, Moad M, Drogo N, Heer R, Sampogna RV, van Rheenen J, Simons BD (2017) A unifying theory of branching morphogenesis. Cell 171:242–255.e27. https://doi.org/10.1016/j.cell.2017.08.026
Hendrix MJC, Seftor EA, Seftor REB, Kasemeier-Kulesa JC, Kulesa PM, Postovit L-M (2007) Reprogramming metastatic tumour cells with embryonic microenvironments. Nat Rev Cancer 7:246–255. https://doi.org/10.1038/nrc2108
Hopkins A, Camley B (2019) Leader cells in collective chemotaxis: optimality and tradeoffs. https://doi.org/10.1101/642157
Iber D, Menshykau D (2013) The control of branching morphogenesis. Open Biol. 3(9). https://doi.org/10.1098/rsob.130088
Inaki M, Vishnu S, Cliffe A, Rørth P (2012) Effective guidance of collective migration based on differences in cell states. Proc Nat Acad Sci 109:2027–2032. https://doi.org/10.1073/pnas.1115260109
Janssens K, Sung H-H, Rørth P (2010) Direct detection of guidance receptor activity during border cell migration. Proc Nat Acad Sci 107:7323–7328. https://doi.org/10.1073/pnas.0915075107
Kasemeier-Kulesa JC, Morrison JA, Lefcort F, Kulesa PM (2015) TrkB/BDNF signalling patterns the sympathetic nervous system. Nat Commun 6:8281. https://doi.org/10.1038/ncomms9281
Kasemeier-Kulesa JC, Romine MH, Morrison JA, Bailey CM, Welch DR, Kulesa PM (2018) NGF reprograms metastatic melanoma to a bipotent glial-melanocyte neural crest-like precursor. Biol Open 7. https://doi.org/10.1242/bio.030817
Keller R (2005) Cell migration during gastrulation. Curr Opin Cell Biol 17:533–541. https://doi.org/10.1016/j.ceb.2005.08.006
Keller EF, Segel LA (1970a) Conflict between positive and negative feedback as an explanation for the initiation of aggregation in Slime Mould Amoebae. Nature 227:1365–1366. https://doi.org/10.1038/2271365a0
Keller EF, Segel LA (1970b) Initiation of slime mold aggregation viewed as an instability. J Theor Biol 26:399–415. https://doi.org/10.1016/0022-5193(70)90092-5
Krotoski DM, Fraser SE, Bronner-Fraser M (1988) Mapping of neural crest pathways in Xenopus laevis using inter- and intra-specific cell markers. Dev Biol 127:119–132. https://doi.org/10.1016/0012-1606(88)90194-7
Kulesa PM, Gammill LS (2010) Neural crest migration: patterns, phases and signals. Dev Biol 344:566–568. https://doi.org/10.1016/j.ydbio.2010.05.005
Kulesa PM, Kasemeier-Kulesa JC, Teddy JM, Margaryan NV, Seftor EA, Seftor REB, Hendrix MJC (2006) Reprogramming metastatic melanoma cells to assume a neural crest cell-like phenotype in an embryonic microenvironment. Proc Nat Acad Sci 103:3752–3757. https://doi.org/10.1073/pnas.0506977103
Kulesa PM, Bailey CM, Kasemeier-Kulesa JC, McLennan R (2010) Cranial neural crest migration: new rules for an old road. Dev Biol 344:543–554. https://doi.org/10.1016/j.ydbio.2010.04.010
La Porta CAM, Zapperi S (2017) The physics of cancer. ISBN:9781316271759
Ladoux B, Mège RM (2017) Mechanobiology of collective cell behaviours. Nat Rev Mol Cell Biol 18:743–757. https://doi.org/10.1038/nrm.2017.98
Lake JI, Heuckeroth RO (2013) Enteric nervous system development: migration, differentiation, and disease. Am J Physiol Gastrointest Liver Physiol 305:G1–24. https://doi.org/10.1152/ajpgi.00452.2012
Lambert B, Maclean AL, Fletcher AG, Coombes AN, Little MH, Byrne HM, Combes AN, Little MH, Byrne HM (2017) Bayesian inference of agent-based models: a tool for studying kidney branching morphogenesis. J Math Biol 0–27. https://doi.org/10.1007/s00285-018-1208-z
Latimer A, Jessen JR (2010) Extracellular matrix assembly and organization during zebrafish gastrulation. Matrix Biol 29:89–96. https://doi.org/10.1016/j.matbio.2009.10.002
Le Douarin NM (2004) Neural crest cell plasticity and its limits. Development 131:4637–4650. https://doi.org/10.1242/dev.01350
Loeb L (1921) Amoeboid movement, tissue formation and consistency of protoplasm. Am J Physiol 56:140–167. https://doi.org/10.1152/ajplegacy.1921.56.1.140
Löfberg J, Ahlfors K, Fällström C (1980) Neural crest cell migration in relation to extracellular matrix organization in the embryonic axolotl trunk. Dev Biol 75:148–167. https://doi.org/10.1016/0012-1606(80)90151-7
Lye CM, Blanchard GB, Naylor HW, Muresan L, Huisken J, Adams RJ, Sanson B (2015) Mechanical coupling between endoderm invagination and axis extension in Drosophila. PLoS Biol 13:e1002292
Maguire LH, Thomas AR, Goldstein AM (2015) Tumors of the neural crest: common themes in development

and cancer. Dev Dyn 244:311–322. https://doi.org/10.1002/dvdy.24226
McKinney MC, Stark DA, Teddy JM, Kulesa PM (2011) Neural crest cell communication involves an exchange of cytoplasmic material through cellular bridges revealed by photoconversion of KikGR. Dev Dyn 240:1391–1401. https://doi.org/10.1002/dvdy.22612
McLennan R, Teddy JM, Kasemeier-Kulesa JC, Romine MH, Kulesa PM (2010) Vascular endothelial growth factor (VEGF) regulates cranial neural crest migration in vivo. Dev Biol 339:114–125. https://doi.org/10.1016/j.ydbio.2009.12.022
McLennan R, Dyson L, Prather KW, Morrison JA, Baker RE, Maini PK, Kulesa PM (2012) Multiscale mechanisms of cell migration during development: theory and experiment. Development 139:2935–2944. https://doi.org/10.1242/dev.081471
McLennan R, Schumacher LJ, Morrison JA, Teddy JM, Ridenour DA, Box AC, Semerad CL, Li H, McDowell W, Kay D, Maini PK, Baker RE, Kulesa PM (2015a) Neural crest migration is driven by a few trailblazer cells with a unique molecular signature narrowly confined to the invasive front. Development 142:2014–2025. https://doi.org/10.1242/dev.117507
McLennan R, Schumacher LJ, Morrison JA, Teddy JM, Ridenour DA, Box AC, Semerad CL, Li H, McDowell W, Kay D, Maini PK, Baker RE, Kulesa PM (2015b) VEGF signals induce trailblazer cell identity that drives neural crest migration. Dev Biol 407:12–25. https://doi.org/10.1016/j.ydbio.2015.08.011
McLennan R, Bailey CM, Schumacher LJ, Teddy JM, Morrison JA, Kasemeier-Kulesa JC, Wolfe LA, Gogol MM, Baker RE, Maini PK, Kulesa PM (2017) DAN (NBL1) promotes collective neural crest migration by restraining uncontrolled invasion. J Cell Biol 216:3339–3354. https://doi.org/10.1083/jcb.201612169
Montell DJ, Yoon WH, Starz-Gaiano M (2012) Group choreography: mechanisms orchestrating the collective movement of border cells. Nat Rev Mol Cell Biol 13:631–645. https://doi.org/10.1038/nrm3433
Muinonen-Martin AJ, Susanto O, Zhang Q, Smethurst E, Faller WJ, Veltman DM, Kalna G, Lindsay C, Bennett DC, Sansom OJ, Herd R, Jones R, Machesky LM, Wakelam MJO, Knecht DA, Insall RH (2014) Melanoma cells break down LPA to establish local gradients that drive chemotactic dispersal. PLOS Biol 12:e1001966. https://doi.org/10.1371/journal.pbio.1001966
Murray JD, Oster GF, Harris AK (1983) A mechanical model for mesenchymal morphogenesis. J Math Biol 17:125–129. https://doi.org/10.1007/BF00276117
Nikitina N, Bronner-Fraser M, Sauka-Spengler T (2009) DiI cell labeling in lamprey embryos. Cold Spring Harb Protoc 4:4–6. https://doi.org/10.1101/pdb.prot5124
Paksa A, Bandemer J, Hoeckendorf B, Razin N, Tarbashevich K, Minina S, Meyen D, Biundo A, Leidel SA, Peyrieras N, Gov NS, Keller PJ, Raz E (2016) Repulsive cues combined with physical barriers and cell-cell adhesion determine progenitor cell positioning during organogenesis. Nat Commun 7:1–14. https://doi.org/10.1038/ncomms11288
Prasad M, Montell DJ (2007) Cellular and molecular mechanisms of border cell migration analyzed using time-lapse live-cell imaging. Dev Cell 12:997–1005. https://doi.org/10.1016/j.devcel.2007.03.021
Prasad M, Wang X, He L, Montell DJ (2011) Border cell migration: a model system for live imaging and genetic analysis of collective cell movement. Methods in molecular biology, vol 769. Humana Press, Totowa, pp 277–286. ISBN:978-1-61779-206-9
Revenu C, Streichan SJ, Donà E, Lecaudey V, Hufnagel L, Gilmour D (2014) Quantitative cell polarity imaging defines leader-to-follower transitions during collective migration and the key role of microtubule-dependent adherens junction formation. Development 141:1282–1291. https://doi.org/10.1242/dev.101675
Reyes M, Zandberg K, Desmawati I, de Bellard ME (2010) Emergence and migration of trunk neural crest cells in a snake, the California Kingsnake (Lampropeltis getula californiae). BMC Dev Biol 10:52. https://doi.org/10.1186/1471-213X-10-52
Ridenour DA, McLennan R, Teddy JM, Semerad CL, Haug JS, Kulesa PM (2014) The neural crest cell cycle is related to phases of migration in the head. Development 141:1095–1103. https://doi.org/10.1242/dev.098855
Rørth P (2012) Fellow travellers: emergent properties of collective cell migration. EMBO Rep 13:984–991. https://doi.org/10.1038/embor.2012.149
Rozbicki E, Chuai M, Karjalainen AI, Song F, Sang HM, Martin R, Knölker H-JJ, MacDonald MP, Weijer CJ (2015) Myosin-II-mediated cell shape changes and cell intercalation contribute to primitive streak formation. Nat Cell Biol 17:397–408. https://doi.org/10.1038/ncb3138
Scarpa E, Mayor R (2016) Collective cell migration in development. J Cell Biol 212:143–155. https://doi.org/10.1083/jcb.201508047
Schilling TF, Kimmel CB (1994) Segment and cell type lineage restrictions during pharyngeal arch development in the zebrafish embryo. Development 120:483–494
Schumacher LJ (2019) Neural crest migration with continuous cell states. J Theor Biol. https://doi.org/10.1016/j.jtbi.2019.01.029
Schumacher LJ, Kulesa PM, McLennan R, Baker RE, Maini PK (2016) Multidisciplinary approaches to understanding collective cell migration in developmental biology. Open Biol 6:160056. https://doi.org/10.1098/rsob.160056
Serbedzija GN, Bronner-Fraser M, Fraser SE (1989) A vital dye analysis of the timing and pathways of avian trunk neural crest cell migration. Development 106:809–816
Serbedzija GN, Fraser S, Bronner-Fraser M (1992) Vital dye analysis of cranial neural crest cell migration in the mouse embryo. Development 116:297–307
Stern CD Gastrulation: from cells to embryo. ISBN:978-087969707-5

Streichan SJ, Valentin G, Gilmour D, Hufnagel L (2011) Collective cell migration guided by dynamically maintained gradients. Phys Biol 8:045004. https://doi.org/10.1088/1478-3975/8/4/045004

Szabó A, Melchionda M, Nastasi G, Woods ML, Campo S, Perris R, Mayor R (2016) In vivo confinement promotes collective migration of neural crest cells. J Cell Biol jcb.201602083. https://doi.org/10.1083/jcb.201602083

Szabó A, Theveneau E, Turan M, Mayor R (2019) Neural crest streaming as an emergent property of tissue interactions during morphogenesis. PLoS Comput Biol 15(4):e1007002. https://doi.org/10.1371/journal.pcbi.1007002

Teddy JM, Kulesa PM (2004) In vivo evidence for short- and long-range cell communication in cranial neural crest cells. Development 131:6141–6151. https://doi.org/10.1242/dev.01534

Tepass U, Fessler LI, Aziz A, Hartenstein V (1994) Embryonic origin of hemocytes and their relationship to cell death in Drosophila. Development (Cambridge, England) 120:1829–1837. https://doi.org/8223268

Theveneau E, Marchant L, Kuriyama S, Gull M, Moepps B, Parsons M, Mayor R (2010) Collective chemotaxis requires contact-dependent cell polarity. Dev Cell 19:39–53. https://doi.org/10.1016/j.devcel.2010.06.012

Trepat X, Sahai E (2018) Mesoscale physical principles of collective cell organization. Nat Phys 1. https://doi.org/10.1038/s41567-018-0194-9

Tweedy L, Knecht DA, Mackay GM, Insall RH (2016) Self-generated chemoattractant gradients: attractant depletion extends the range and robustness of Chemotaxis. PLOS Biol 14:e1002404. https://doi.org/10.1371/journal.pbio.1002404

Weijer CJ (2009) Collective cell migration in development. J Cell Sci 122:3215–3223. https://doi.org/10.1242/jcs.036517

Wolpert L (2008) The triumph of the Embryo. ISBN:0486469298

Nuclear Mechanics and Cancer Cell Migration

8

Charlotte R. Pfeifer, Jerome Irianto, and Dennis E. Discher

Abstract

As a cancer cell invades adjacent tissue, penetrates a basement membrane barrier, or squeezes into a blood capillary, its nucleus can be greatly constricted. Here, we examine: (1) the passive and active deformation of the nucleus during 3D migration; (2) the nuclear structures—namely, the lamina and chromatin—that govern nuclear deformability; (3) the effect of large nuclear deformation on DNA and nuclear factors; and (4) the downstream consequences of mechanically stressing the nucleus. We focus especially on recent studies showing that constricted migration causes nuclear envelope rupture and excess DNA damage, leading to cell cycle suppression, possibly cell death, and ultimately it seems to heritable genomic variation. We first review the latest understanding of nuclear dynamics during cell migration, and then explore the functional effects of nuclear deformation, especially in relation to genome integrity and potentially cancerous mutations.

C. R. Pfeifer · J. Irianto · D. E. Discher (✉)
Biophysical Engineering Labs: Molecular & Cell Biophysics and NanoBio-Polymers, University of Pennsylvania, Philadelphia, PA, USA
e-mail: discher@seas.upenn.edu

Keywords

Nuclear mechanics · Constricted migration · Nuclear lamina rupture · DNA damage · Cell cycle · Genome instability

8.1 Introduction

Multiple processes *in vivo* require cells to move through three-dimensional (3D) tissue. Cancer cells migrate into wounds during healing (Clark et al. 1982) and into vessel-adjacent matrix during angiogenesis (Lamalice et al. 2007). Leukocytes squeeze through capillaries as small as 2–3 μm in diameter and extravasate into sites of tissue damage or infection as part of the innate immune response (Luster et al. 2005). Embryogenesis involves progenitor and committed cells moving and positioning themselves in developing organs (Kurosaka and Kashina 2008). Cancer cells invade healthy tissue, penetrate basement membrane barriers, and enter distant capillary beds during tumor metastasis (Liotta et al. 1991). As the largest and stiffest organelle (Dahl et al. 2008), the nucleus has long been speculated to sterically limit a cell's ability to migrate through small, stiff pores including basement membranes that separate tissues (Lichtman 1970). In migration through constricting 3D fibrous matrix, the nucleus has been described as a "piston" that is pulled forward to establish a hydrostatic pressure

C. A. M. La Porta, S. Zapperi (eds.), *Cell Migrations: Causes and Functions*, Advances in Experimental Medicine and Biology 1146,
https://doi.org/10.1007/978-3-030-17593-1_8

gradient between the tight-fitting nucleus and the cell's leading edge (Petrie et al. 2014). Moreover, softening the nucleus by knockdown of key nuclear structure components can enhance the rate of migration through small constrictions (Shin et al. 2013; Harada et al. 2014), consistent with the idea of the nucleus as a physical impediment to migration.

While the nucleus affects migration by presenting a challenge to the moving cell, migration also affects the nucleus. Constriction-induced deformation causes chromatin reorganization and even nuclear envelope rupture (Denais et al. 2016; Raab et al. 2016; Irianto et al. 2016a, 2017), among other effects, with implications for important biological processes like DNA damage and repair. For an overall understanding of cell migration, it is therefore necessary to consider the role of the nucleus.

The chapter will examine: (1) the forces exerted on the nucleus during 3D migration; (2) the regulators of nuclear deformability that influence transit through small pores; (3) the impact of large nuclear deformation on chromatin and nuclear factors; and (4) the downstream consequences of physically perturbing the nuclear content, including effects on genome integrity and cell cycle progression. A main goal of the chapter is to introduce some of the biophysical processes relevant to nuclear dynamics during cell migration, while also highlighting the functional effects of nuclear deformation on the biology of the cell.

8.2 Structure of the Nucleus

Although nuclear sizes vary among and even within cell types, the nucleus is typically the largest cellular organelle, with a diameter of ~5–20 μm (Dahl et al. 2008). In cells imaged *in situ* or grown in 3D scaffolds, the nucleus tends to be round or ovoid, whereas 2D culture drives cell spreading and nuclear flattening (Khatau et al. 2009). The nucleus—along with other, smaller organelles (<1–2 μm) and cytoskeletal filaments—is embedded in the cell's gel-like cytoplasm. The cytoplasm, cytoskeleton, and plasma membrane are easily deformed and rearranged during constricted migration such that cytoplasmic protrusions can squeeze into channels of even submicron diameter (Wolf et al. 2013). By contrast, the nucleus is 2–10 times stiffer than the surrounding cell body (Guilak et al. 2000; Caille et al. 2002), making its constriction a more torturous—and rate-limiting (Davidson et al. 2015)—step in the process of 3D migration.

The nuclear envelope, which defines the boundary of the nucleus, consists of two closely apposed lipid bilayers: the inner nuclear membrane (INM) and the outer nuclear membrane (ONM). Both are continuous with the endoplasmic reticulum. Just below the INM is the nuclear lamina (Fig. 8.1), a dense meshwork of intermediate filament proteins (lamins) that confers mechanical support and stiffness to the nuclear envelope (Ungricht and Kutay 2017). Together, the envelope and lamina surround the nucleoplasm, the genome (i.e. chromatin), and various subnuclear bodies—mostly ribonucleic protein complexes like nucleoli, promyelocytic leukemia (PML) nuclear bodies, Cajal bodies, and splicing speckles (Martins et al. 2012).

The nucleus mechanically couples to the cytoskeleton by way of Linker of Nucleoskeleton and Cytoskeleton (LINC) complexes (Irianto et al. 2016b). A LINC complex consists of a SUN protein that binds to the lamina and connects *via* nuclear envelope spectrin repeat proteins (nesprins) to a KASH domain on the ONM (Tapley and Starr 2013). The cytoplasmic region of the KASH domain then mediates interactions between the nucleus and the cytoplasm/cytoskeleton by tethering the ONM to cytoskeletal microtubules, actin filaments, and intermediate filaments (Tapley and Starr 2013). Numerous experiments demonstrate this physical nucleo-cytoskeletal linkage: for example, targeted laser ablation of the actin cytoskeleton causes the nucleus to move laterally and away from the culture substrate, and can even cause local nuclear deformation (Mazumder and Shivashankar 2010; Nagayama et al. 2011). And the disabling of endogenous LINC complexes results in loss of cellular mechanical stiffness comparable to the loss of stiffness observed with lam-

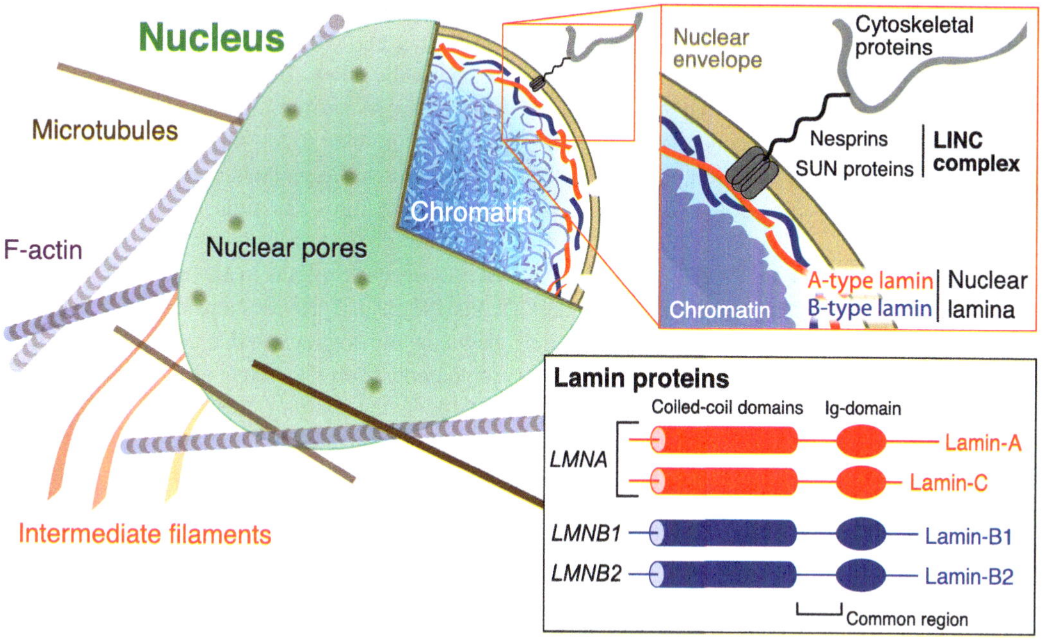

Fig. 8.1 A-type and B-type lamins form a dense meshwork on the inside of the nuclear envelope. The nucleus mechanically couples to the cytoskeleton by way of Linker of Nucleoskeleton and Cytoskeleton (LINC) complexes, which consist of a SUN protein that binds to the lamina and connects *via* nuclear envelope spectrin repeat proteins (nesprins) to a KASH domain on the outer nuclear membrane. The cytoplasmic region of the KASH domain then mediates interactions between the nucleus and the cytoplasm/cytoskeleton by tethering the nuclear membrane to cytoskeletal microtubules, actin filaments, and intermediate filaments

ina disruption (Stewart-Hutchinson et al. 2008). Because the nucleus mechanically couples to the cytoskeleton—and ultimately, via adhesions, to extracellular matrix (ECM)—it deforms with the cell during 3D migration (Broers et al. 2004; Swift et al. 2013).

8.3 Deformation of the Nucleus During Constricted Migration

Different cell types employ different single-cell migration modes in 3D environments. Tissue fibroblasts exhibit relatively slow (~0.5–1 μm/minute) mesenchymal cell migration (Cukierman et al. 2001), while dendritic cells and immune cells (e.g. leukocytes) favor ~10–40-fold faster amoeboid movement (Friedl et al. 1998). Mesenchymal cell migration proceeds as follows: (1) the cell polarizes to create a leading edge that extends actin-rich protrusions; (2) the protrusions form adhesions to ECM contacts; (3) myosin II-mediated contraction of the actin cytoskeleton shortens the rear of the cell and advances the cell body; and (4) the trailing edge detaches from the substratum, allowing the cell to translate forward. Such migration can include degradation of the ECM by matrix matelloproteinases (MMPs) (Even-Ram and Yamada 2005). By contrast, amoeboid migration is typically non-proteolytic, and it involves weaker, more transient adhesive interactions with the ECM (Parri and Chiarugi 2010). In reality, migration modes exist along a continuum, and the mode adopted by a given cell in a given microenvironment seems to be determined by factors such as ECM stiffness and the intrinsic contractility of the cell (Parsons et al. 2010).

Regardless of the particular motility mode, deformation of the cell during 3D migration leads to deformation of the nucleus. The first step in the migration process, polarization, requires the cell's cytoskeleton and organelles—including the nucleus—to rearrange themselves within the cell body. In polarized mesenchymal cells, fibroblasts, neurons, and most cancer cells, the nucleus assumes a rearward position, thus establishing a leading edge-to-centrosome-to-nucleus axis along the direction of locomotion—at least on rigid substrates (Gomes et al. 2005; Gasser and Hatten 1990; Tsai and Gleeson 2005; Barnhart et al. 2010). On soft substrates, the centrosome is more random in location (Raab et al. 2012). Whereas most nuclear movements are thought to be microtubule-mediated (Luxton et al. 2010), nuclear repositioning for migration is driven by retrograde flow of actin: inhibiting myosin-II or actin with blebbistatin or cytochalasin D, respectively, is known to block actin retrograde flow, and doing so prevents rearward nuclear movement during cell polarization (Gomes et al. 2005), although cells can still migrate with myosin-II inhibition in 2D. Actin cables are coupled to the dorsal surface of the nuclear envelope by LINC complexes, as described above; these linkages transmit force from actin flow to the nucleus (Luxton et al. 2010).

After polarization, as the cell proceeds to squeeze into a tight constriction in 3D, the nucleus moves with the cell body by being either pushed or pulled. Under the pushing mechanism, the nucleus is squeezed forward by actomyosin contraction in the (detached) rear of the cell (Roth et al. 1995; Zhang et al. 2007). Such trailing-edge contraction propels nuclear translocation during constricted migration of leukocytes: myosin II-inhibited leukocytes migrating through 3D gels exhibit a peculiar elongated shape with a rounded back due to nuclear immobilization at the rear ends of the cells. Because posterior actomyosin contraction is required to retract and detach the cell membrane, myosin II inhibition renders leukocytes unable to push their large, rigid nuclei through small interstices in the gel (Lammermann et al. 2008). Similarly, in 3D migration studies of breast, brain, and other cancer cells, non-muscle myosin II localizes to the perinuclear cytoskeleton and cell posterior, and then exerts pushing forces to advance the nucleus (Harada et al. 2014; Beadle et al. 2008; Ivkovic et al. 2012). Knockdown of myosin IIB dramatically slows migration of breast cancer cells through narrow channels, whereas knockdown of myosin IIA—the non-muscle myosin II isoform that generates force during leading-edge protrusion—has little effect on migration time through the constrictions. The isoforms have almost opposite effects in 3D migration of glioma cells. Nesprin-2 provides a possible physical link between the nucleus and myosin IIB-mediated force generation (Beadle et al. 2008).

Under the pulling mechanism, actomyosin contraction physically pulls the nucleus forward during 3D migration. When Rac1 photoactivation is used to create a new leading protrusion in a crawling fibroblast (by triggering local F-actin polymerization at the front of the cell), the nucleus moves persistently toward the new leading edge without trailing-edge detachment—even when microtubules are depolymerized (Wu et al. 2014). In lobopodial fibroblasts, the pulling forces are generated by non-muscle myosin IIA-containing actomyosin bundles that form complexes with the intermediate filament protein vimentin and the LINC protein nesprin-3 (Petrie et al. 2014). Ultimately, it is likely that both pushing and pulling forces contribute—in a cell type- and migration mode-dependent manner—to the forward motion of nuclei during 3D migration.

Whether the nucleus is pushed or pulled by actomyosin, it can undergo huge deformation when constricted. Whereas the nucleus maintains its original ellipsoid shape and simply re-orients during transit through large pores in loose tissues (Friedl et al. 2011), it is severely locally compressed by small pores in dense tissues, resulting in transient shape changes (Harada et al. 2014). Reflecting the larger deformation required by smaller pores, migration speed decreases linearly with decreasing pore size (Irianto et al. 2017). Compression of the nucleus during migration is actuated by cytoskeletal forces and opposed by the geometry of ECM pores. In 2D culture on stiff glass substrates, a dome-like perinuclear actin cap largely aligns with the overall cell orientation (Khatau et al. 2009), and this cap might actively drive nuclear shape changes during 3D migration. Moreover, intermediate filaments including vimentin surround the nucleus in a fibrous "cage" that is required for nuclear re-shaping in response to actomyosin-induced forces (Neelam et al. 2015). Both the actin cap and cage-like intermediate filaments connect to the nuclear envelope through LINC complexes; hence, LINC complex disruption impairs nucleo-cytoskeletal-mediated nuclear deformation and often causes migratory defects (Khatau et al. 2012).

Beyond single-cell migration, it should be noted that cells often maintain their cell-cell junctions and undergo collective migration, traveling in sheets, strands, tubes, or clusters (Parri and Chiarugi 2010). Such movement usually occurs along smooth ECM interfaces (Friedl et al. 2011); for example, collective migration of invasive cancer cells through tissue barriers requires MMPs to clear tracks—devoid of sterically impeding fibers—in the ECM. Multicellular invasion along these proteolytic tracks causes significantly less morphological adaptation and nuclear deformation than does single-cell migration through non-reorganized collagen (Wolf et al. 2007). Thus, the severity of nuclear deformation depends on the mode of migration—collective versus single-cell, proteolytic versus non-proteolytic.

8.4 Regulators of Nuclear Deformability

Nuclei have viscoelastic properties (Guilak et al. 2000; Dahl et al. 2005), meaning that they exhibit stress relaxation: when a constant deformation is applied, the resulting mechanical stress on the nucleus decays over time. They also exhibit a creep response such that when a constant stress is applied, the nucleus continues permanently to deform. Viscoelastic materials are often modeled as a network of elastic springs and viscous dashpots. For example, in the three-component standard linear solid model, which is designed to show exponential stress relaxation and exponential creep, a spring is placed in parallel with a "Maxwell arm" consisting of a spring and dashpot in series (Meidav 1964). This model has been applied to isolated articular chondrocyte nuclei pulled by constant suction pressure into micropipettes (Guilak et al. 2000). However, spring-dashpot models are limited in the case of nuclei because nuclear stress relaxation and creep occur over many decades of time. To accurately model viscoelastic behavior on such timescales would require a very large (physically meaningless) number of spring and dashpot elements, which could increase mathematical complexity to the point of impracticality (Lange and Fabry 2013).

As opposed to a superposition of very many exponential response functions, a power-law model provides a simpler and more physically meaningful approach to describe nuclear mechanics under deformation. Indeed, micropipette aspiration and atomic force microscopy (AFM) assays indicate that isolated intestinal epithelial cell nuclei exhibit power-law rheology (Dahl et al. 2005). The creep compliance $J(t)$ of the nucleus—that is, the ratio of nuclear strain to applied stress as a function of time t—is given by

$$J(t) = J_0\left(\frac{t}{\text{sec}}\right)^{\alpha} [=] \frac{1}{\text{kPa}}, \tag{8.1}$$

where the prefactor J_0 corresponds to the inverse of the dynamic shear modulus G measured at a frequency of 1 rad/s (Dahl et al. 2005; Lange and Fabry 2013; Hildebrandt 1969). The exponent α depends on the dynamics of the force-bearing elastic structures of the nucleus (Lange and Fabry 2013)—in particular, the lamina and the chromatin, as we will describe below. A purely elastic solid would have a power-law exponent of $\alpha = 0$, while a purely viscous fluid would have an exponent of $\alpha = 1$. The measured value for isolated nuclei from intestinal epithelial cells is $\alpha \approx 0.2$–0.3 (Dahl et al. 2005).

Power-law rheology could have a number of important consequences for nuclei undergoing constricted migration. First, in a material with a power-law exponent of $\alpha \approx 0.2$–0.3, mechanical stresses decay slower than exponentially, but they do become small for large enough t. To illustrate, if the effective stiffness of a nucleus is 1 kPa when measured at a frequency of 1 Hz, then the same nucleus should have an effective stiffness of only ∼0.3 kPa when measured at 0.01 Hz. Thus, ignoring active mechanics, as the speed of nuclear movement decreases, so do the movement-resisting forces that arise from nuclear deformation (Lange and Fabry 2013). Second, the power-law behavior of nuclei has implications for chromatin organization during migration. Other systems with power-law rheology, such as microgels, have an essentially infinite number of intermediate conformations corresponding to infinite relaxation modes or timescales. It seems likely that nuclear components at different length scales—from nucleosomes to chromosomes to chromatin fibers—also have intermediate conformations of mechanical relaxation, reflecting metastable states that could impact gene expression kinetics (Dahl et al. 2005).

The power-law viscoelasticity of nuclei is determined principally by the lamina and chromatin—or at least the chromatin volume fraction. We will discuss each structure in turn. The intermediate filaments that comprise the nuclear lamina are divided into two sub-types (Fig. 8.1): A-type lamins (lamin-A and -C), which are alternative splicing products of the LMNA gene; and B-type lamins (lamin-B), which are encoded by the LMNB1 and LMNB2 genes. Although A- and B-type lamins have similar amino acid sequences and structural features, they have different post-translational modifications (Irianto et al. 2016b): the lamin-B monomer is permanently modified by addition of a farnesyl group, which is hydrophobic and tethers lamin-B to the INM (Hennekes and Nigg 1994). As a result, lamin-B is less mobile and dynamic than mature lamin-A (Shimi et al. 2008), from which the farnesylated C-terminus is cleaved (Irianto et al. 2016b). Like other intermediate filament proteins, including keratin and vimentin, lamin monomers form coiled-coil parallel dimers that assemble into filaments of ∼3.5 nm thickness, organized in complex meshworks of ∼14 nm thickness (Herrmann et al. 2009; Turgay et al. 2017). Cryo-electron tomography of mouse embryonic fibroblasts suggests that both lamin sub-types are present throughout the meshwork, including in densely packed and sparsely occupied regions (Turgay et al. 2017).

Lamin-A levels vary widely across adult cell types, scaling with resident tissue stiffness (Swift et al. 2013). Meanwhile, lamin-B expression remains relatively constant such that the ratio of lamin-A to -B is highest in stiff tissues like muscle and bone, and lowest in soft tissues like brain and fat. The positive scaling of lamin-A:B ratio with tissue microelasticity suggests a possible role for lamin-A in protecting the nucleus against mechanical stresses, which are expected to be higher in stiffer tissues. Consistent with such a protective function, lamin-A confers viscous stiffness to nuclei, while lamin-B contributes to nuclear elasticity. When nuclei of diverse tissue lineage are pulled into micropipettes under controlled pressure (∼kPa), each nucleus extends within seconds in a viscoelastic manner, as described above. Importantly, effective nuclear viscosity increases more rapidly than effective elasticity as a function of lamin-A:B stoichiometry. This trend suggests that whereas lamin-B functions like the elastic walls of a balloon, restoring the nucleus to its original shape in response to applied stresses, lamin-A acts like a viscous fluid that coats the walls and perhaps fills the balloon to dynamically resist deformation (Swift et al. 2013). Moreover, lamin-A knockdown is known to soften nuclei (Harada et al. 2014; Pajerowski et al. 2007), and mutations in lamin-A are associated with diseases—"laminopathies" including muscular dystrophy and premature aging (Sullivan et al. 1999). Levels of lamin-A and lamin-B are abnormal in many cancers; lamin-A is low in lung and breast tumors, for example (Irianto et al. 2016b). Lamin-A depletion had been reported to favor nuclear rupture in fibroblastic cells spread and flattened on stiff substrates but not on soft substrates where cells and nuclei are more rounded and relaxed (Tamiello et al. 2013); further study of such 2D cultures demonstrated that nuclear rupture occurs at sites of high nuclear curvature where stiff lamin-B filaments tend to detach (Xia et al. 2018).

In cell migration through small pores, the lamina regulates nuclear deformability. Lamin-A in particular is known to be rate-limiting in 3D migration of diverse human cell lines, ranging from brain and lung cancer cells to primary mesenchymal stem cells (MSCs) (Harada et al. 2014). For a given cell type, wild-type levels of lamin-A protect against stress-induced death during transit through small pores, whereas low levels cause susceptibility to stress and apoptosis, and high levels impede migration. Thus, lamin-A is a barrier to 3D migration, but it promotes nuclear integrity and survival (Harada et al. 2014).

Chromatin can also play a role in the mechanical response of the nucleus. Chromatin consists of DNA wrapped around histone octamers, and it exists in two forms: open euchromatin (low density), which contains most actively transcribed genes; and tightly packed heterochromatin (higher density), which can silence gene transcription (Dahl et al. 2008). Treatment with the deacetylase inhibitor trichostatin A (TSA) favors euchromatin organization by causing large-scale chromatin decondensation; such remodeling renders nuclei softer and more deformable. Conversely, chromatin condensation by divalent cations such as Ca^{2+} and Mg^{2+} results in extremely stiff nuclei with small values for the creep compliance factors J_0 and α (Eq. 8.1). The relative deformability of euchromatin structures hints that external forces—like those imposed by constricting pores during 3D migration—could easily reorganize gene-rich regions of the genome (Pajerowski et al. 2007). Moreover, although isolated chromosomes respond elastically to applied stress (Cui and Bustamante 2000; Marko 2008), chromatin within the nucleus responds by either flowing with the stress or bearing it. As observed in nuclei that lack lamins, INM chromatin tethers constrain flow and favor an elastic response to small forces. However, untethering of chromatin from the INM allows the chromatin to flow under deformation to a new, lower energy configuration (Schreiner et al. 2015). These results, which indicate that chromatin contributes both elasticity and viscosity to the nucleus, are consistent with: (1) nuclear stretching experiments in which chromatin governs elastic resistance to small nuclear deformations (Stephens et al. 2017); and (2) micropipette aspiration experiments showing that chromatin can flow, shear, and locally compact like a complex fluid (Pajerowski et al. 2007). Together, the lamina and the chromatin (especially in case of high volume fraction) determine the mechanical properties of the nucleus and the severity of nuclear deformation during constricted migration.

8.5 Effect of Nuclear Deformation on Chromatin and Nuclear Factors

Migration through constricting pores exerts compressive forces on the nucleoplasm, causing the internal pressure in the nucleus to rise. In regions—like the leading tip of the nucleus—where no external forces are applied, the increased internal pressure is equilibrated by an increase in the surface tension of the nuclear envelope, per the Young-Laplace equation. To relax some of this tension and lower the membrane stretching energy, a hole may form in the lamina, leading to fluid outflow from the nucleus that locally inflates the nuclear envelope. Such inflation produces a bleb, which can burst to cause leakage of nuclear factors—and even herniation of chromatin—into the cytoplasm, with corresponding leakage of cytoplasmic factors into the nucleus (Deviri et al. 2017). Indeed, in back-to-back papers from two groups (Denais et al. 2016; Raab et al. 2016), migration of various cancer cell lines, immortalized epithelial cells, and primary dendritic cells through narrow channels was shown to rupture the nuclear envelope. Rupture, which can occur even without bleb formation (Pfeifer et al. 2018), leads to exchange of nucleo-cytoplasmic contents, as indicated by cytoplasmic accumulation of GFP-NLS (nuclear localization signal) and nuclear accumulation of NES (nuclear export signal)-GFP. The resealing of nuclear envelope lesions is thought to be mediated by endosomal sorting complex required for transport III (ESCRT III) components (Denais et al. 2016; Raab et al. 2016).

In the same two papers, rupture was shown to be followed by enrichment at the envelope (Denais et al. 2016) or far from the envelope (Raab et al. 2016) of a GFP fusion of an overexpressed DNA repair factor 53BP1, but no supporting evidence of DNA damage was provided in terms of endogenous damage markers such as the standard histone γH2AX or electrophoresis of DNA fragments. The authors speculated that the observed pools of GFP-53BP1 could be due to nuclear influx of cytoplasmic nucleases, which potentially cleave the DNA and trigger a DNA damage response. However, accumulations of GFP-53BP1 could instead reflect local pooling of mobile nuclear proteins into rarefied pockets of the strongly deformed chromatin (Irianto et al. 2016a).

Other pore migration studies of two cancer lines and primary human MSCs have provided clear measures of excess DNA damage based on increased foci of γH2AX, increased foci of the upstream kinase phospho-ATM, and longer electrophoretic comets of DNA (Irianto et al. 2017; Pfeifer et al. 2018). Nuclear entry of cytoplasmic nucleases fails as an explanation for this damage, because nuclease infiltration would be expected to cause localized damage concentrated near the site of nuclear envelope rupture. However, γH2AX and pATM foci have a pan-nucleoplasmic distribution, suggesting a more global damage mechanism (Irianto et al. 2017). While it is tempting to propose that chromatin fragmentation as a nucleus enters and elongates in a small pore might account for the increased damage, this mechanism also seems unlikely given that stretched chromatin maintains its integrity. In living cells, an mCherry-tagged nuclease was targeted to a submicron locus on chromosome 1, where it causes DNA cleavage—and thus recruitment of DNA repair factors to a large region around the locus. Micropipette aspiration of these cells and their nuclei shows that the chromatin aligns and stretches parallel to the pore axis. Importantly, even though DNA within the engineered locus is cleaved by nuclease, intensity profiles of mCherry indicate continuity, meaning that integrity of the chromatin is maintained during nuclear distention. Chromatin shearing is therefore unlikely to explain the excess DNA damage that follows pore migration (Irianto et al. 2016c).

One possible mechanism is global inhibition of DNA repair. DNA breaks constantly form by various means, including replication or oxidative stress, and are repaired by dedicated factors that are often implicated in cancer (including ATM, BRCA1, etc.). Damage rate and repair rate reach a steady state dependent on the level or activity of the repair factors. A compelling hypothesis holds that constriction-induced mis-localization of repair factors causes partial depletion of repair factors throughout the nucleus, which physically inhibits repair of routine DNA breaks and leads to the observed transient increases in DNA damage (Irianto et al. 2017; Pfeifer et al. 2018). Inactivating mutations in DNA repair factors BRCA1 and BRCA2 are such well-established risk factors for cancer that they warrant surgical removal of ovary and breasts (Levy-Lahad and Friedman 2007). Mouse knockouts or heterozygous mutants for BRCA1 and BRCA2, among other repair proteins, have indeed been shown to alter chromosome copy numbers (Holstege et al. 2010). Therefore, any migration-induced physical depletion of such factors should also increase DNA damage and mutation probabilities.

Constriction mis-localizes repair factors in two ways (Fig. 8.2). First, if nuclear envelope rupture occurs, as described above, then diffusible repair factors leak into the cytoplasm for hours before ultimately re-localizing to the nucleus (Irianto et al. 2017). Second, regardless of nuclear envelope rupture, constrictions "squeeze out" all diffusible proteins from regions of high DNA compaction, such as at the entrance of a constricting pore (Irianto et al. 2016a). To elaborate, absent DNA damage, GFP-53BP1 is ordinarily diffuse in the nucleus, consistent with nucleoplasmic mobility (Bekker-Jensen et al. 2005; Pryde et al. 2005). However, during constricted migration, as the nucleus contorts to enter a pore, mobile GFP-53BP1 is significantly depleted within the

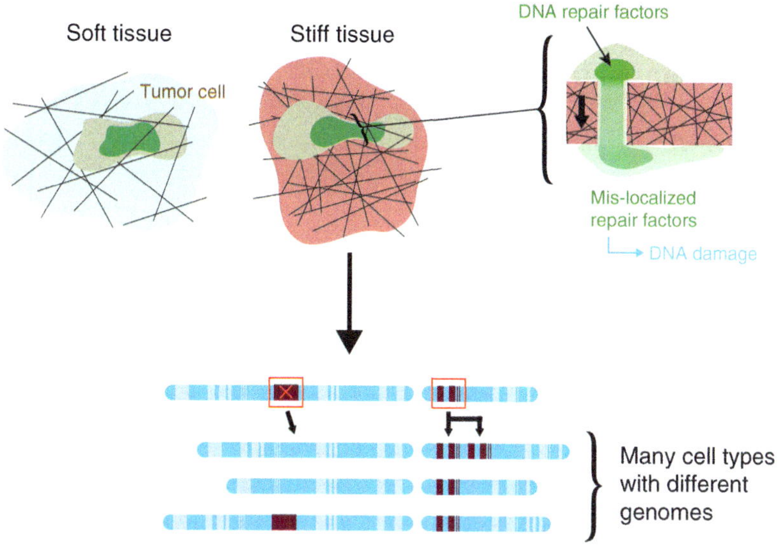

Fig. 8.2 Since stiffer tissues tend to have higher matrix density, and thus smaller pores, cancer cells might sustain more nuclear stress during tumorigenic invasion into stiff tissues as compared to soft ones. Migration through small pores severely deforms the nuclei of invading cancer cells, which causes mis-localization of DNA repair factors *via* (1) "squeeze-out" of all diffusible proteins from regions of high DNA compaction and (2) rupture-induced leakage of nuclear proteins into the cytoplasm. Such mis-localization causes partial depletion of repair factors throughout the nucleus, which physically inhibits repair of routine DNA breaks and leads to excess DNA damage. Migration-induced DNA damage results in lasting, heritable genomic heterogeneity

constriction, in contrast to DNA or chromatin-bound proteins like mCherry-Histone H2B, which are instead enriched in the constriction. Similarly, endogenous 53BP1 (immunostained) and the additional DNA repair factors GFP-Ku70 and -Ku80—all in the mobile phase—show such striking depletion. These observations suggest that nuclear constriction excludes, and hence depletes, mobile nucleoplasmic factors from the pore. Nuclear factor segregation is also observed during micropipette aspiration: all of a dozen mobile proteins examined—including upstream DNA damage response factors (e.g. MRE11, RPA) as well as downstream factors (e.g. BRCA1)—segregate like GFP-53BP1 (Irianto et al. 2016a).

A simple model for squeeze-out of mobile nuclear factors provides insight into why segregation occurs and gives a mechanistic basis for the hypothesis that severe constriction can arrest DNA damage repair (Bennett et al. 2017). Chromatin is modeled as a solid mesh of volume fraction f, intermixed with a fluid of mobile nuclear proteins. For cells in static culture, chromatin has been measured to occupy $f \sim 67\%$ of the nuclear volume (Bancaud et al. 2009), so the free volume for diffusion of mobile factors is $(1 - f) \sim 33\%$. However, constriction increases the local density of chromatin by a factor of $\sim$1.25 such that inside the pore $f_{\text{constricted}} \sim 84\%$, which causes the free volume there to decrease to $(1 - f_{\text{constricted}}) \sim 16\%$. It follows that mobile factors should decline in the constriction to $(1 - f_{\text{constricted}}) / (1 - f) \sim 0.5$ of their original abundance, which agrees well with experiments (Irianto et al. 2016a; Bennett et al. 2017). This depletion of mobile proteins, including repair proteins, is compounded by rupture-induced leakage into the cytoplasm. Altogether, mis-localization of repair factors during constricted migration impedes the DNA damage response. Thus, normally occurring DNA damage—that might arise due to replication stress, reactive oxygen species, or other sources—cannot be efficiently repaired while the nucleus is inside the pore, leading to excess damage.

8.6 Downstream Consequences of Constriction-Induced Nuclear Deformation

Migration-induced DNA damage leads to lasting genomic heterogeneity that translates to transcriptomic and phenotypic changes. Clonal U2OS cells were subjected to three consecutive migrations through constricting pores; from among these thrice-migrated cells, the genomes of six single-cell-derived clones were quantified by SNP array analysis. Compared to the pre-migration clone, the migrated clones showed unique chromosome copy number changes and loss of heterozygosity (Irianto et al. 2017). Because the pre-migration population was 100% clonal and the migrated sub-clones exhibited unique genomic changes, it stands to reason that migration causes—as opposed to simply selecting for—genomic variation. On the other hand, selection of rare subpopulations with pre-existing genomic differences is a possible mechanism that might only be addressed by live cell monitoring of genome transitions in single cells – a method that awaits development.

The clinical implications of constricted migration causing heritable mutations are vast. Advances in whole-genome sequencing technology have allowed for complete cataloguing of the genomic changes that occur in cancers of different types (Martin et al. 2015; Schumacher and Schreiber 2015; Matsushita et al. 2016). In a meta-analysis of published cancer sequencing data (Pfeifer et al. 2017), the somatic mutation rates for 36 cancer types were culled from a number of recent papers (Schumacher and Schreiber 2015; Alexandrov et al. 2013; Lawrence et al. 2013; Martincorena et al. 2015; Martincorena and Campbell 2015; Chen et al. 2014; Shain et al. 2015), as were the stiffnesses of the healthy tissues in which those cancers arise. This meta-analysis revealed that cancers arising in stiff tissues, such as lung and skin, exhibit more than 30-fold higher somatic mutation rates than those arising in soft tissues, like marrow and brain. Although tumors often stiffen—or, less frequently, soften—their surrounding tissue over the course of tumorigenesis (Levental et al. 2010), the stiffness of a typical brain tumor microenvironment never reaches that of a typical bone tumor microenvironment, so the stiffness gradient among tissue types prevails.

The scaling of genomic variation with tissue stiffness suggests a possible mechanical source of cancerous mutations. One promising hypothesis implicates constricted migration of cells through stiff tissues (Fig. 8.2). Tissue stiffness increases with abundance of fibrous protein (e.g. collagen) (Swift et al. 2013), and denser collagen matrix has smaller interstitial pores (Yang et al. 2009). Therefore, when cancer cells invade normal tissue during tumor growth (Liotta et al. 1991), they generally encounter a higher collagen concentration and smaller pores in stiffer tissues than in softer ones. As discussed, squeezing through small pores—but not larger ones—severely deforms the nuclei of invading cancer cells, which stresses the nuclear lamina and causes DNA damage, heritable genome changes, and even cell death (Harada et al. 2014; Irianto et al. 2017). Thus, constricted migration through increasingly small holes in increasingly collagen-rich matrix stands as a possible explanation for the relation between mutation rate and tissue stiffness.

In a recent study, single-cell genome sequencing was used to measure copy number changes in single breast tumor cells while preserving their spatial context in the breast tissue. The authors of this study found a direct genomic lineage between the primary tumor (the ductal carcinoma *in situ*) and invasive tumor subpopulations; they concluded that the subpopulations must carry mutations from the primary carcinoma, rather than incurring new mutations during the invasive migration process (Casasent et al. 2018). However, breast is of low-to-intermediate stiffness (Lopez et al. 2011), so it makes sense, per the above hypothesis, that migration of tumor cells through breast tissue does not cause a large increase in mutational load.

Since cells need to repair DNA damage sufficiently in order to progress through cell cycle (Dasika et al. 1999), it seems plausible that DNA breaks incurred during migration—perhaps in combination with mis-localization of crucial cell

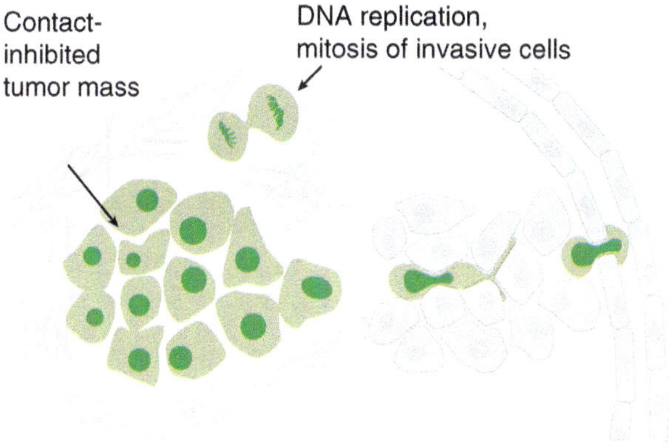

Fig. 8.3 Invasion and proliferation are hallmarks of cancer. Invading cancer cells squeeze into regions of low cancer cell density, including nearby tissues or blood capillaries. The resulting loss of contact inhibition could in principle encourage proliferation. However, migration through 3 μm pores has been shown to cause a transient delay in cell cycle for diverse cancer cell lines, illustrating a "go, damage, and then grow" process

cycle proteins—could suppress cell proliferation. Combined EdU cell proliferation and pore migration assays show that migration through 3 μm pores indeed causes a transient delay in cell cycle for three diverse cancer cell lines (Pfeifer et al. 2018). These findings are relevant to the so-called "go or grow" hypothesis, long-debated in cancer research, which holds that proliferation and migration are mutually exclusive events (Garay et al. 2013; Giese et al. 1996). It appears that additional mutation-relevant processes are involved in a "go, damage, *repair*, and then grow" behavior, with cancer cells showing excess DNA damage and repressed cell cycle after migration. The surprising delay in growth has implications for the invasive migration of cancer cells away from a physically crowded tumor mass and into nearby stiff tissues or blood capillaries (Fig. 8.3). Moreover, the combined proliferation/migration assays also show that G1- and G2-phase cells incur a similar excess of DNA damage, suggesting that constriction-induced DNA damage occurs independent of cell cycle phase and hence independent of DNA replication (Pfeifer et al. 2018).

8.7 Conclusion

In this chapter, we began by describing the deformation of the nucleus during 3D migration, both due to forces imposed by the geometry of rigid pores and due to intracellular mechanisms that actively drive nuclear shape change. Then, we reviewed the latest research on the mechanical properties of the nucleus and, in particular, examined how the lamina and chromatin regulate nuclear deformability during migration through tight constrictions. Next, we considered the impact of such large deformation on chromatin and nuclear factors. Constricted migration causes frequent lamina rupture, which—along with 'squeeze-out' of mobile nuclear proteins—leads to mis-localization of crucial DNA repair factors, followed by an increase in DNA damage. Finally, we discussed some downstream consequences of constriction-induced DNA damage, namely, effects on genome integrity including possible mutations, as well as cell death and delays in cell cycle progression. Differentiation is also seen to be affected based on studies to be published soon. This chapter has introduced biophysics concepts relevant to nuclear mechanics during cell migration, while outlining some of the biological consequences of severe nuclear deformation.

References

Alexandrov LB, Nik-Zainal S, Wedge DC, Aparicio SAJR, Behjati S, Biankin AV, Bignell GR, Bolli N, Borg A, Borresen-Dale A-L et al (2013) Signatures of mutational processes in human cancer. Nature 500:415–421

Bancaud A, Huet S, Daigle N, Mozziconacci J, Beaudouin J, Ellenberg J (2009) Molecular crowding affects diffusion and binding of nuclear proteins in heterochromatin and reveals the fractal organization of chromatin. EMBO J 28:3785–3798

Barnhart EL, Allen GM, Julicher F, Theriot JA (2010) Bipedal locomotion in crawling cells. Biophys J 98:933–942

Beadle C, Assanah MC, Monzo P, Vallee R, Rosenfeld SS, Canoll P (2008) The role of myosin II in glioma invasion of the brain. Mol Biol Cell 19(8):3357–3368. https://doi.org/10.1091/mbc.E08-03-0319

Bekker-Jensen S, Lukas C, Melander F, Bartek J, Lukas J (2005) Dynamic assembly and sustained retention of 53BP1 at the sites of DNA damage are controlled by Mdc1/NFBD1. J Cell Biol 170:201–211

Bennett RR, Pfeifer CR, Irianto J, Xia Y, Discher DE, Liu AJ (2017) Elastic-fluid model for DNA damage and mutation from nuclear fluid segregation due to cell migration. Biophys J 112(11):2271–2279

Broers JLV, Peeters EAG, Kuijpers HJH, Endert J, Bouten CVC, Oomens CWJ, Baaijens FPT, Ramaekers FCS (2004) Decreased mechanical stiffness in LMNA –/– cells is caused by defective nucleo-cytoskeletal integrity: implications for the development of laminopathies. Hum Mol Genet 13:2567–2580

Caille N, Thoumine O, Tardy Y, Meister J-J (2002) Contribution of the nucleus to the mechanical properties of endothelial cells. J Biomech 35:177–187

Casasent AK, Schalck A, Gao R, Sei E, Long A, Pangburn W, Casasent T, Meric-Bernstam F, Edgerton ME, Navin NE (2018) Multiclonal invasion in breast tumors identified by topographic single cell sequencing. Cell 172:205–217

Chen X, Bahrami A, Pappo A, Easton J, Dalton J, Hedlund E, Ellison D, Shurtleff S, Wu G, Wei L et al (2014) Recurrent somatic structural variations contribute to tumorigenesis in pediatric osteosarcoma. Cell Rep 7:104–112

Clark RAF, Lanigan JM, DellaPelle P, Manseau E, Dvorak HF, Colvin RP (1982) Fibronectin and fibrin provide a provisional matrix for epidermal cell migration during wound reepithelialization one-micron section studies of the migrating epidermis. J Invest Dermatol 79:264–269

Cui Y, Bustamante C (2000) Pulling a single chromatin fiber reveals the forces that maintain its higher-order structure. PNAS 97:127–132

Cukierman E, Pankov R, Stevens DR, Yamada KM (2001) Taking cell-matrix adhesions to the third dimension. Science 294:1708–1713

Dahl KN, Engler AJ, Pajerowski JD, Discher DE (2005) Power-law rheology of isolated nuclei with deformation mapping of nuclear substructures. Biophys J 89:2855–2864

Dahl KN, Ribeiro AJS, Lammerding J (2008) Nuclear shape, mechanics, and mechanotransduction. Circ Res 102:1307–1319

Dasika GK, Lin S-CJ, Zhao S, Sung P, Tomkinson A, Lee EY-HP (1999) DNA damage-induced cell cycle checkpoints and DNA strand break repair in development and tumorigenesis. Oncogene 18:7883–7899

Davidson PM, Denais C, Bakshi MC, Lammerding J (2015) Nuclear deformability constitutes a rate-limiting step during cell migration in 3-D environments. Cell Mol Bioeng 7:293–306

Denais CM, Gilbert RM, Isermann P, Mcgregor AL, te Lindert M, Weigelin B, Davidson PM, Friedl P, Wolf K, Lammerding J (2016) Nuclear envelope rupture and repair during cancer cell migration. Science 352:353–358

Deviri D, Discher DE, Safran SA (2017) Rupture dynamics and chromatin herniation in deformed nuclei. Biophys J 113:1060–1071

Even-Ram S, Yamada KM (2005) Cell migration in 3D matrix. Curr Opin Cell Biol 17:524–532

Friedl P, Zanker KS, Brocker E-B (1998) Cell migration strategies in 3-D extracellular matrix: differences in morphology, cell matrix interactions, and integrin function. Microsc Res Tech 43:369–378

Friedl P, Wolf K, Lammerding J (2011) Nuclear mechanics during cell migration. Curr Opin Cell Biol 23:55–64

Garay T, Juhász É, Molnár E, Eisenbauer M, Czirók A, Dekan B, László V, Alireza Hoda M, Döme B, Tímár J et al (2013) Cell migration or cytokinesis and proliferation? – revisiting the "go or grow" hypothesis in cancer cells in vitro. Exp Cell Res 319: 3094–3103

Gasser UE, Hatten ME (1990) Central nervous system neurons migrate on astroglial fibers from heterotypic brain regions in vitro. PNAS 87:4543–4547

Giese A, Loo MA, Tra N, Haskett D, Coons SW, Berens ME (1996) Dichotomy of astrocytoma migration and proliferation. Int J Cancer 67:275–282

Gomes ER, Jani S, Gundersen GG (2005) Nuclear movement regulated by Cdc42, MRCK, myosin, and actin flow establishes MTOC polarization in migrating cells. Cell 121:451–463

Guilak F, Tedrow JR, Burgkart R (2000) Viscoelastic properties of the cell nucleus. Biochem Biophys Res Commun 269:781–786

Harada T, Swift J, Irianto J, Shin JW, Spinler KR, Athirasala A, Diegmiller R, Dingal PCDP, Ivanovska IL, Discher DE (2014) Nuclear lamin stiffness is a barrier to 3D migration, but softness can limit survival. J Cell Biol 204:669–682

Hennekes H, Nigg EA (1994) The role of isoprenylation in membrane attachment of nuclear lamins. J Cell Sci 107:1019–1029

Herrmann H, Strelkov SV, Burkhard P, Aebi U (2009) Intermediate filaments: primary determinants of cell architecture and plasticity. J Clin Invest 119:1772–1783

Hildebrandt J (1969) Institute of respiratory physiology, virginia mason research center and firland sanatorium, and the department of physiology and biophysics. Bull Math Biophys 31:651–667

Holstege H, Horlings HM, Velds A, Langerød A, Børresen-Dale A-L, Van De Vijver MJ, Nederlof PM, Jonkers J (2010) BRCA1-mutated and basal-like breast

cancers have similar aCGH profiles and a high incidence of protein truncating TP53 mutations. BMC Cancer 10:455–470
Irianto J, Pfeifer CR, Bennett RR, Xia Y, Ivanovska IL, Liu AJ, Greenberg RA, Discher DE (2016a) Nuclear constriction segregates mobile nuclear proteins away from chromatin. Mol Biol Cell 27:4011–4020
Irianto J, Pfeifer CR, Ivanovska IL, Swift J, Discher DE (2016b) Nuclear lamins in cancer. Cell Mol Bioeng 9:258–267
Irianto J, Xia Y, Pfeifer CR, Greenberg RA, Discher DE (2016c) As a nucleus enters a small pore, chromatin stretches and maintains integrity, even with DNA breaks. Biophys J 112:446–449
Irianto J, Xia Y, Pfeifer CR, Athirasala A, Ji J, Alvey C, Tewari M, Bennett RR, Harding SM, Liu AJ et al (2017) DNA damage follows repair factor depletion and portends genome variation in cancer cells after pore migration. Curr Biol 27:210–223
Ivkovic S, Beadle C, Noticewala S, Massey SC, Swanson KR, Toro LN, Bresnick AR, Canoli P, Rosenfeld SS (2012) Direct inhibition of myosin II effectively blocks glioma invasion in the presence of multiple motogens. Mol Biol Cell 23:533–542
Khatau SB, Hale CM, Stewart-Hutchinson PJ, Patel MS, Stewart CL, Searson PC, Hodzic D, Wirtz D (2009) A perinuclear actin cap regulates nuclear shape. PNAS 106:19017–19022
Khatau SB, Bloom RJ, Bajpai S, Razafsky D, Zang S, Giri A, Wu P-H, Marchand J, Celedon A, Hale CM et al (2012) The distinct roles of the nucleus and nucleus-cytoskeleton connections in three-dimensional cell migration. Sci Rep 2:488
Kurosaka S, Kashina A (2008) Cell biology of embryonic migration. Birth Defects Res C Embryo Today 84:102–122
Lamalice L, Le Boeuf F, Huot J (2007) Endothelial cell migration during angiogenesis. Circ Res 100:782–794
Lammermann T, Bader BL, Monkley SJ, Worbs T, Wedlich-Soldner R, Hirsch K, Keller M, Forster R, Critchley DR, Fassler R et al (2008) Rapid leukocyte migration by integrin- independent flowing and squeezing. Nature 453:51–55
Lange JR, Fabry B (2013) Cell and tissue mechanics in cell migration. Exp Cell Res 319:2418–2423
Lawrence MS, Stojanov P, Polak P, Kryukov GV, Cibulskis K, Sivachenko A, Carter SL, Stewart C, Mermel CH, Roberts SA et al (2013) Mutational heterogeneity in cancer and the search for new cancer-associated genes. Nature 499:214–218
Levental KR, Yu H, Kass L, Lakins JN, Egeblad M, Erler JT, Fong SFT, Csiszar K, Giaccia A, Weninger W et al (2010) Matrix crosslinking forces tumor progression by enhancing integrin signaling. Cell 139: 891–906
Levy-Lahad E, Friedman E (2007) Cancer risks among BRCA1 and BRCA2 mutation carriers. Br J Cancer 96:11–15
Lichtman MA (1970) Cellular deformability during maturation of the myeloblast. N Engl J Med 283:943–948
Liotta LA, Steeg PS, Stetler-Stevenson WG (1991) Cancer metastasis and angiogenesis: an imbalance of positive and negative regulation. Cell 64:327–336
Lopez JI, Kang I, You W-K, McDonald DM, Weaver VM (2011) In situ force mapping of mammary gland transformation. Integr Biol 3:910–921
Luster AD, Alon R, von Andrian UH (2005) Immune cell migration in inflammation : present and future therapeutic targets. Nat Immunol 6:1182–1190
Luxton GWG, Gomes ER, Folker ES, Vintinner E, Gundersen GG (2010) Linear arrays of nuclear envelope proteins harness retrograde actin flow for nuclear movement. Science 329:956–960
Marko JF (2008) Micromechanical studies of mitotic chromosomes. Chromosom Res 16:469–497
Martin SD, Coukos G, Holt RA, Nelson BH (2015) Targeting the undruggable: immunotherapy meets personalized oncology in the genomic era. Ann Oncol 26:2367–2374
Martincorena I, Campbell PJ (2015) Somatic mutation in cancer and normal cells. Science 349:961–968
Martincorena I, Roshan A, Gerstung M, Ellis P, Van LP, Mclaren S, Wedge DC, Fullam A, Alexandrov LB, Tubio JM et al (2015) High burden and pervasive positive selection of somatic mutations in normal human skin. Science 348:880–886
Martins RP, Finan JD, Guilak F, Lee DA (2012) Mechanical regulation of nuclear structure and function. Annu Rev Biomed Eng 14:431–455
Matsushita H, Sato Y, Karasaki T, Nakagawa T, Kume H, Ogawa S, Homma Y, Kakimi K (2016) Neoantigen load, antigen presentation machinery, and immune signatures determine prognosis in clear cell renal cell carcinoma. Cancer Immunol Res 4:463–471
Mazumder A, Shivashankar GV (2010) Emergence of a prestressed eukaryotic nucleus during cellular differentiation and development. J R Soc Interface 7:S321–S330
Meidav T (1964) Viscoelastic properties of the standard linear solid. Geophys Prospect 12:1365–2478
Nagayama K, Yahiro Y, Matsumoto T (2011) Stress fibers stabilize the position of intranuclear DNA through mechanical connection with the nucleus in vascular smooth muscle cells. FEBS Lett 585:3992–3997
Neelam S, Chancellor TJ, Li Y, Nickerson JA, Roux KJ, Dickinson RB, Lele TP (2015) Direct force probe reveals the mechanics of nuclear homeostasis in the mammalian cell. PNAS 112:5720–5725
Pajerowski JD, Dahl KN, Zhong FL, Sammak PJ, Discher DE (2007) Physical plasticity of the nucleus in stem cell differentiation. PNAS 104:15619–15624
Parri M, Chiarugi P (2010) Rac and Rho GTPases in cancer cell motility control. Cell Commun Signal 8:1–14
Parsons JT, Horwitz AR, Schwartz MA (2010) Cell adhesion: integrating cytoskeletal dynamics and cellular tension. Nat Rev Mol Cell Biol 11:633–643
Petrie RJ, Koo H, Yamada KM (2014) Generation of compartmentalized pressure by a nuclear piston governs cell motility in 3D matrix. Science 345:1062–1065

Pfeifer CR, Alvey CM, Irianto J, Discher DE (2017) Genome variation across cancers scales with tissue stiffness – An invasion-mutation mechanism and implications for immune cell infiltration. Curr Opin Syst Biol 2:102–113

Pfeifer CR, Xia Y, Zhu K, Liu D, Irianto J, García VMM, Millán LMS, Niese B, Harding S, Deviri D et al (2018) Constricted migration increases DNA damage and independently represses cell cycle. Mol Biol Cell. https://doi.org/10.1091/mbc.E18-02-0079

Pryde F, Khalili S, Robertson K, Selfridge J, Ritchie A-M, Melton DW, Jullien D, Adachi Y (2005) 53BP1 exchanges slowly at the sites of DNA damage and appears to require RNA for its association with chromatin. J Cell Sci 118:2043–2055

Raab M, Swift J, Dingal PDP, Shah P, Shin J-W, Discher DE (2012) Crawling from soft to stiff matrix polarizes the cytoskeleton and phosphoregulates myosin-II heavy chain. J Cell Biol 199:669–683

Raab M, Gentili M, de Belly H, Thiam HR, Vargas P, Jimenez AJ, Lautenschlaeger F, Voituriez R, Lennon-Dumenil AM, Manel N et al (2016) ESCRT III repairs nuclear envelope ruptures during cell migration to limit DNA damage and cell death. Science 352:359–362

Roth S, Neuman-silberberg FS, Barcelo G, Schlipbach T (1995) Cornichon and the EGF receptor signaling process are necessary for both anterior-posterior and dorsal-ventral pattern formation in drosophila. Cell 81:967–978

Schreiner SM, Koo PK, Zhao Y, Mochrie SGJ, King MC (2015) The tethering of chromatin to the nuclear envelope supports nuclear mechanics. Nat Commun 6:7159

Schumacher TN, Schreiber RD (2015) Neoantigens in cancer immunotherapy. Science 348:69–74

Shain AH, Yeh I, Kovalyshyn I, Sriharan A, Talevich E, Gagnon A, Dummer R, North J, Pincus L, Ruben B et al (2015) The genetic evolution of melanoma from precursor lesions. N Engl J Med 373:1926–1936

Shimi T, Pfleghaar K, Kojima S, Pack C-G, Solovei I, Goldman AE, Adam SA, Shumaker DK, Kinjo M, Cremer T et al (2008) The A- and B-type nuclear lamin networks: microdomains involved in chromatin organization and transcription. Genes Dev 22:3409–3421

Shin J-W, Spinler KR, Swift J, Chasis JA, Mohandas N, Discher DE (2013) Lamins regulate cell trafficking and lineage maturation of adult human hematopoietic cells. PNAS 110:18892–18897

Stephens AD, Banigan EJ, Adam SA, Goldman RD, Dunn AR (2017) Chromatin and lamin A determine two different mechanical response regimes of the cell nucleus. Mol Biol Cell. https://doi.org/10.1091/mbc.E16-09-0653

Stewart-Hutchinson PJ, Hale CM, Wirtz D, Hodzic D (2008) Structural requirementsfor the assembly of LINC complexes and their function in cellular mechanical stiffness. Exp Cell Res 314:1892–1905

Sullivan T, Escalante-Alcalde D, Bhatt H, Anver M, Bhat N, Nagashima K, Stewart CL, Burke B (1999) Loss of A-type lamin expression compromises nuclear envelope integrity leading to muscular dystrophy. J Cell Biol 147:913–919

Swift J, Ivanovska IL, Buxboim A, Harada T, Dingal PCDP, Pinter J, Pajerowski JD, Spinler KR, Shin J, Tewari M et al (2013) Nuclear lamin-A scales with tissue stiffness and enhances matrix-directed differentiation. Science 341:1240104

Tamiello C, Kamps MA, van den Wijngaard A, Verstraeten VL, Baaijens FP, Broers JL, Bouten CC (2013) Soft substrates normalize nuclear morphology and prevent nuclear rupture in fibroblasts from a laminopathy patient with compound heterozygous LMNA mutations. Nucleus 4:61–73

Tapley EC, Starr DA (2013) Connecting the nucleus to the cytoskeleton by SUN – KASH bridges across the nuclear envelope. Curr Opin Cell Biol 25:57–62

Tsai L-H, Gleeson JG (2005) Nucleokinesis in neuronal migration minireview. Neuron 46:383–388

Turgay Y, Eibauer M, Goldman AE, Shimi T, Khayat M, Ben-Harush K, Dubrovsky-Gaupp A, Sapra KT, Goldman RD, Medalia O (2017) The molecular architecture of lamins in somatic cells. Nature 543:261–264

Ungricht R, Kutay U (2017) Mechanisms and functions of nuclear envelope remodelling. Nat Rev Mol Cell Biol 18:229–245

Wolf K, Wu YI, Liu Y, Geiger J, Tam E, Overall C, Stack MS, Friedl P (2007) Multi-step pericellular proteolysis controls the transition from individual to collective cancer cell invasion. Nat Cell Biol 9:893–904

Wolf K, te Lindert M, Krause M, Alexander S, te Riet J, Willis AL, Hoffman RM, Figdor CG, Weiss SJ, Friedl P (2013) Physical limits of cell migration: control by ECM space and nuclear deformation and tuning by proteolysis and traction force. J Cell Biol 201:1069–1084

Wu J, Kent IA, Shekhar N, Chancellor TJ, Mendonca A, Dickinson RB, Lele TP (2014) Actomyosin pulls to advance the nucleus in a migrating tissue cell. Biophys J 106:7–15

Xia Y, Ivanovska IL, Zhu K, Smith L, Irianto J, Pfeifer CR, Alvey CM, Ji J, Liu D, Cho S, Bennett RR, Liu AJ, Greenberg RA, Discher DE (2018) Nuclear rupture at sites of high curvature compromises retention of DNA repair factors. J Cell Biol 217:3796–3808

Yang Y, Leone LM, Kaufman LJ (2009) Elastic moduli of collagen gels can be predicted from two-dimensional confocal microscopy. Biophys J 97:2051–2060

Zhang X, Xu R, Zhu B, Yang X, Ding X, Duan S, Xu T, Zhuang Y, Han M (2007) Syne-1 and Syne-2 play crucial roles in myonuclear anchorage and motor neuron innervation. Development 134:901–908

Index

C. A. M. La Porta, S. Zapperi (eds.), *Cell Migrations: Causes and Functions*, Advances in Experimental Medicine and Biology 1146,
https://doi.org/10.1007/978-3-030-17593-1

The manufacturer's authorised representative in the EU is Springer Nature Customer Service Centre GmbH, Europaplatz 3, 69115 Heidelberg, Germany. If you have any concerns regarding our products, please contact ProductSafety@springernature.com

Printed and bound by CPI Group (UK) Ltd, Croydon, CR0 4YY
28/09/2026
02224079-0007

ORGANIC FARMING
Scope and Use of Biofertilizers

NIPA® GENX ELECTRONIC RESOURCES & SOLUTIONS P. LTD.
New Delhi-110 034